Software User Documentation

A How To Guide for Project Staff

David Tuffley

To my beloved Nation of Four
Concordia Domi – Foris Pax

Wise men speak because they have something to say;
Fools because they have to say something. Plato

Acknowledgements

I am indebted to the Institute of Electrical and Electronics Engineers on whose work I base this book, specifically IEEE Std 1063.

I also acknowledge the *Turrbal* and *Jagera* indigenous peoples, on whose ancestral land I write this book.

Contents

A.   **Introduction** **6**

   A.1.   *User manuals are not technical* 7

   A.2.   *Why we need good user documentation* 7

   A.3.   *Manuals written in the early stages* 8

   A.4.   *On-line manuals / help files* 9

   A.5.   *Practices what it preaches* 9

B.   **Preparation** **10**

   B.1.   *Source material* *11*

   B.2.   *Documentation project plan* *12*

   B.3.   *Project plan* *15*

      B.3.1.   Document name (section 1) 15

      B.3.2.   Purpose (section 2) 15

         B.3.2.1.   Instructional mode 16

         B.3.2.2.   Reference mode 17

      B.3.3.   Table of contents (section 3) 17

      B.3.4.   Deliverables (section 4) 17

      B.3.5.   Resources (section 5) 18

      B.3.6.   Nature of the software (section 6) 18

      B.3.7.   Scope (section 7) 18

      B.3.8.   Topic (section 8) 19

      B.3.9.   Who is the audience? (section 9) 19

      B.3.10.   Document set (section 10) 20

      B.3.11.   Presentation (section 11) 21

      B.3.12.   Project schedule (section 12) 21

      B.3.13.   Software (section 13) 22

      B.3.14.   Editing (section 14) 22

      B.3.15.   Budget (section 15) 22

      B.3.16.   Copyright (section 16) 22

      B.3.17.   Translation (section 17) 23

Contents

B.4.   Estimating *23*

C.   The writing process **26**

C.1.   Clear & effective communication *26*

C.2.   No tired figures of speech *27*

C.3.   Short not long words *27*
C.3.1.   Economical & precise with words 29
C.3.2.   Active not passive 30
C.3.3.   Everyday English not foreign, jargon or scientific 31
C.3.4.   Prefabricated language 31
C.3.5.   Present tense not past/future 32
C.3.6.   Avoiding overstatement 32
C.3.7.   Adapting words to the reader 33
C.3.8.   Never barbarous (advisory only) 33

C.4.   Non sexist language *35*

C.5.   Writer's block *35*
C.5.1.   Preparation 36
C.5.2.   Make a start 36
C.5.3.   Review the reference material 37

C.6.   Environment *37*

C.7.   Routine *38*

C.8.   Ergonomics *39*
C.8.1.   Chair 39
C.8.2.   Screen 39
C.8.3.   Regular breaks 40
C.8.4.   Keyboard 40

D.   The first draft **41**

D.1.   Backups *41*

D.2.   Deciding on the required user documents *42*

D.3.   User document inclusion requirements *43*

D.4.   User document content requirements *45*
D.4.1.   Title page 45
D.4.2.   Inside page 46

Contents

D.4.3.	Table of contents	47
D.4.3.1.	Comprehensive table of contents	48
D.4.3.2.	Simple table of contents	48
D.4.4.	List of Illustrations	48
D.4.5.	Headers/footers	49
D.4.6.	Document Introduction	50
D.4.6.1.	Audience description	50
D.4.6.2.	Applicability statement	51
D.4.6.3.	Purpose statement	51
D.4.6.4.	Document usage description	51
D.4.6.5.	Related documents	52
D.4.6.6.	Conventions	52
D.4.6.7.	Problem solving instructions	53
D.4.7.	Body of document	53
D.4.7.1.	Body of instructional documents	54
D.4.7.2.	Body of reference documents	56
D.4.8.	Error messages, known problems & error recovery	59
D.4.9.	Appendixes	59
D.4.9.1.	Bibliography	60
D.4.10.	Glossary	61
D.4.11.	Index	61
D.4.12.	User documentation presentation requirements	62
D.4.12.1.	Consistency	62
D.4.12.2.	Terminology	62
D.4.12.3.	Referencing related material	63
D.4.13.	Change control	63
D.4.13.1.	This-version function changes	63
D.4.13.2.	Next-version function changes	63
E.	**Document production**	**64**
E.1.	*Document orientation*	*64*
E.2.	*Document organisation*	*65*
E.3.	*Document format*	*65*
E.3.1.	Double sided	65
E.3.2.	Table of contents	66
E.3.3.	Page size and margins	67
E.3.4.	Headings	67
E.3.5.	Body text	68
E.3.6.	Bullets	69

Contents

E.3.7. Representing screens.................................... 70
   E.3.7.1. Overview 70
   E.3.7.2. Screen capture techniques........................ 71
   E.3.7.3. Screen requirements.............................. 72
E.3.8. Keyboard function keys 72
E.3.9. Numbered lists 73
E.3.10. Figures & numerals................................. 73
E.3.11. Justification 73
E.3.12. Italics.. 74
E.3.13. Emphasis 74
E.3.14. Hyphenation.. 75
E.3.15. Acronyms & jargon.................................. 75
E.3.16. Widow/orphan protection............................ 76
E.3.17. Paragraph numbering................................ 76
E.3.18. Other considerations 76

F. Editing & proof-reading 77

F.1. Editing 77
F.1.1. Print first draft.................................. 77
F.1.2. Binding.. 78
F.1.3. Edit first draft................................... 78
F.1.4. Edit second draft 80
F.1.5. Spell/grammar check, TOC, index.................... 80
F.1.6. First proof.. 80
F.1.7. Checklist 81

F.2. Proof-reading 82
F.2.1. Who proof-reads?................................... 83
F.2.2. Deadlines.. 84
   F.2.2.1. Make the changes 85

G. Reviewing & field testing............................... 86

G.1. Reviewing the manual................................. 86
G.1.1. Selecting reviewers and when to review 87
G.1.2. Show reviewers how to review 88
G.1.3. Give feedback to reviewers......................... 89

G.2. Field testing the manual 89
G.2.1. Selecting testers and when to test................. 90
G.2.2. Show testers how to test........................... 90
G.2.3. Give feedback to testers 91

Contents

H. Production 92

   H.1. Paper size & weight 92

   H.2. Print quality 93

   H.3. Binding 93

   H.4. Drilling 93

   H.5. Tab dividers 94

   H.6. Registration 94

   H.7. Identification of changes 96

   H.8. Document release 96

   H.9. Version control & re-issue of documents 97

   H.10. Controlled distribution documents 98

I. On-line documentation 99

   I.1. Conversion to on-line 100

   I.2. Create the paper manual first 100

J. Maintenance 101

   J.1. When to maintain 101

   J.2. Procedure 101

   J.3. Change control 102

      J.3.1. Post-publication software changes 102

      J.3.2. Post-publication document changes 103

K. Glossary of computer terms 104

L. Bibliography 135

M. Appendix A - Forms 137

A. Introduction

User documentation is basically about finding the best way of putting across technical information in a non-technical way to non-technical people. If we want to do the greatest amount of good for the largest number of computer users, then it is the end-user documentation audience who require the most attention, since they are the largest consumers of computer documentation.

User documentation is the most difficult type of communication for IT people (*"techos"*) to do well. This is because user documentation calls for communication between people with widely different backgrounds.

User documentation does not necessarily require writing that simplifies the software. User documentation is best thought of as a type of writing that *translates* the computer activities for the benefit of non-technical users and readers. User documentation therefore takes data processing information and translates into ideas that can be readily understood by people who are skilled in other areas or other disciplines.

User documentation presents a situation much like that involved with translating a foreign language. When you are speaking to someone who shares a common background and language, they can fill in the gaps and make up for mistakes in your communication. However when speaking to someone who has a different background and speaks a different language, they cannot make up for gaps and mistakes, and need additional explanation.

A.1. User manuals are not technical

Think of the difference between a programmer's reference guide and the user guide that comes with common software packages. The first contains highly technical language and concepts, while the second is expressed in the kind of language the user is likely to understand.

Whereas technical writing catalogues facts in detail, user documentation should only say as much as is necessary to describe a step by step process. Too often, a document intended for user consumption is confused with technical writing, resulting in a document that is doomed to failure unless the user is willing and able to think like a technician.

A.2. Why we need good user documentation

Without a doubt the computer industry suffers from a lack of good user documentation and it is the user who suffers. More often than not the software is written and then someone says, "Oh, and of course we'll need a manual". So, taking the technical specifications that the programmers used, someone puts it through a word processor and it comes out the other side as the "user manual". The problem is that it still sounds like a technical document. The user takes one look at it, thinks "This is useless" and after that the manual stays on the shelf gathering dust until it falls victim to a cleanup some years later.

On the other hand, large software companies spend a lot on their manuals in order for them to be "friendly" to the average user. That is to say, presented so that the user can easily find the information they need to get on with the job at hand.

The same principles which the big software companies use to produce those manuals are given in this book. If you follow the steps outlined in this guide, your manual will be useful to the reader.

A.3. Manuals written in the early stages

As far back as the 1980's the then powerful Digital Equipment Corporation (DEC) noted in their internal documentation guidelines that user documentation should be written first - not last as is traditionally done - because the user documentation is an excellent way to debug the design of a system or program. If a writer finds it difficult to document a system, the problem is probably the system, not the writer. Holes in design, obscure constructions and apparent contradictions become starkly visible in the documentation.

Potential problem areas can be identified early rather than after the product is finished. If problems are identified early, they are more likely to be fixed, since fixing problems later in the development cycle costs more money. In fact the later a problem is found, the more costly it becomes to fix.

Most software products that can be described as being outstandingly good and/or user friendly have been developed this way.

A.4. On-line manuals / help files

There are a range of proprietary on-line manual / help file authoring tools available on the market if the reader needs them. The quality and functionality of these tools varies. With the changing landscape of this market, this book makes no recommendations. You can choose the package that best meets your needs, based on your requirements.

But even if you will be only be producing on-line and not paper-based manuals, you still have to know *what* to put into your online manual, and *how* to organise the information for best effect. The information in this book is still useful for this.

So if you are developing on-line help, use this guide to write the manual, the content and organisation of which can then be fed into the help authoring tool, and/or printed as a paper-based manual.

A.5. Practices what it preaches

Many of the ideas and principles outlined by this "how to" guide are illustrated by the document itself. So if you are in doubt as to what is intended, look at the way this document does it.

B.　Preparation

This chapter covers the range of activities that are necessary to establish a good foundation for your documentation project. The importance of good preparation can hardly be emphasised enough - it is vital to the success of any documentation project, particularly larger projects.

Preparation is another name for planning, and as the old saying goes, *failing to plan is planning to fail*. The Return on Investment (ROI) in software development projects generally has been estimated at around 6:1. For every hour spent planning, six hours are saved or gained later in the project.

The object is to clearly define the finished product before beginning to write. If you do this, and get approval for your concept so that the project manager and yourself both expect the same thing, you will avoid the need to re-do sections of the manual later.

Much trouble can be avoided by having a clear agreement between yourself and the project manager as to what the finished product will look like and contain. Unfounded assumptions must be avoided since they are usually not shared by the parties.

In a sense, good planning is about creating a shared vision of what the future product, in this case a manual, will look like. As every project manager knows, one of the hardest things about running projects is managing stakeholder expectations. Good planning does just that.

B.1. Source material

It is important that all source material necessary for a thorough treatment of the subject be made available to you at the outset.

Source material includes the following:

- All relevant specifications, record formats and screen and report layouts.
- An operating copy of the software, if available.]
- The analysts and programmers of the software, including the timely resolution of questions raised by you.
- Where available, typical users for audience analysis and usability testing.

While it is your responsibility to communicate accurately the source material to the user, you are not personally responsible for the accuracy of the original source material.

Where documentation standards already exist, you need to be supplied with copies of the relevant standards. Otherwise, use the standard outlined in a later section of this book.

Source material received by you is normally on a loan basis. You need to keep it in good order and return it when finished. In some cases the information might be confidential. In these cases, you will need to sign a non-disclosure agreement.

B.2. Documentation project plan

It is necessary to develop a project plan. This is done in two stages. The first stage is to be briefed by the project manager or someone delegated by them as to the exact nature of the job. The results of the briefing are a mutual understanding of what is to be delivered, and by when.

Prepare a document called *"Preliminary Project Plan"* using the headings which follow. This will form the basis of the final project plan which can't be developed until the following information is gathered and considered.

- **Title -** working title, purpose, scope and limitations of document.
- **Audience** - who will read the manual?
- **Why used** - reasons why the documentation would be used by the intended audience, and for what purpose.
- **Table of contents** - draft table of contents.
- **Deliverables** - how many printed copies (whether on CD/DVD or printed) are to be supplied, disk and file formats (including software versions) and where and when they will be delivered.
- **People -** who and what resources will be available to help you? What person hours are required? At what cost?
- **Resources** - project resource requirements, including the information and other resources that the customer will have to provide and when.

- **Change control** - a specified method for passing information about software changes to the documenter during the development of the manual.
- **Milestones** - a schedule showing appropriate milestones, including where appropriate:
- Documentation plan approval.
- Preparation, review and approval of each draft.
  - Usability testing.
  - Camera-ready artwork preparation.
  - Printing, binding and distribution.
- **Source material** - what written information - i.e. requirements list, specifications, reports, etc. is available?
- **Software** - is the software to be documented available for you to use?
- **Standards/specifications** - is the documentation to be written to a particular standard or specification? - for example a company style sheet or standard format. The "Document production" section of this manual is an example.
- **Technical edit** - who will check the technical accuracy of the manual?
- **Editorial control** - who has editorial control?
- **Budget** - how much will it cost?
- **Copyright** - ownership of copyright and any other proprietary rights. Usually, the author of a commissioned document is the first owner unless a specific agreement otherwise. If the customer is to own

the copyright, it must be assigned as part of the contract between author and customer.

- **Translation** - provision for the translation into other languages, if applicable.

Getting answers to these questions allows you to develop a suitable project plan.

When you have finished the preliminary project plan, discuss it with the project manager and after any amendments which might arise from those discussions, it should be signed-off or otherwise formally agreed by both parties. This might seem an unnecessary formality however from experience its value as a benchmark for future reference is undoubted. It also obliges the project manager to provide you with everything you need to complete the project - this sometimes comes in very handy.

B.3. Project plan

Once the preliminary project plan is complete and it has been signed-off, a detailed project plan is developed.

Using the preliminary plan as a basis, develop the project plan using the following headings shown below.

Note: a series of forms to support the following activities are provided in appendix A.

B.3.1. Document name (section 1)

Decide on a name for the manual. Be guided by the subject matter and the intended audience. Be as concise as possible when describing the purpose. A manual can also be called a handbook or a guide.

For example, "CMS User Guide" is more concise than "Complete Guide to the Construction Management System".

B.3.2. Purpose (section 2)

A clear statement of the purpose of the document. This will start with what you put in the preliminary plan, together with any amendments which might have arisen from your discussions with the project manager.

Specify the usage mode of the document. The usage mode of any user documentation can be divided into two broad categories - instructional and reference:

- **Instructional** - where the user needs to learn about the software. Instructional documents are either informational (i.e. tutorials and introductory manuals) or task-oriented (i.e. quick reference manuals which give the steps the user needs to perform a task).

- **Reference** - where the user needs to refer to information or refresh their memory and which give the information needed to understand the subject.

It's important to decide the usage mode at the outset. Without this definition, the manual will probably be a mixture, alternating between modes, which will confuse the reader. On the other hand, the impression of *consistency* inspires confidence in the document - something you very much want if the manual is to succeed.

B.3.2.1. Instructional mode

An instructional mode document should give the background information needed to understand the system, and also give details of the range of functions the software can perform and how to perform them. Examples are given to reinforce the learning process.

The Informational type of instructional document provides background and any technical detail needed to understand the system. This would include a system overview, a theory of operation and a tutorial. The technical information should be provided by the people who developed the system.

Task oriented instructional documents outline the procedures of operation. Examples of this kind of documentation include a diagnostic procedures manual (i.e. troubleshooting manual), operations manual and software installation manual (getting started or read this first).

B.3.2.2. Reference mode

Reference mode documents organises information that users might need, and allows quick access to details of specific subjects. Examples of reference mode documents include a command manual, error messages, program calls, quick reference guide, software tools and utilities.

B.3.3. Table of contents (section 3)

Outline the draft table of contents.

B.3.4. Deliverables (section 4)

Specify how many printed copies are required, whether disk copies are to be supplied, the disk and file formats (including software versions) and where and when they will be delivered.

B.3.5. Resources (section 5)

Specify the project resource requirements, including what information is required from who and when. Who and what resources are to be made available?

What source material, written information is to be made available. This includes requirements list, specifications, reports etc.

B.3.6. Nature of the software (section 6)

What is the nature of the software? Is a working copy available for you to use? For example is it for accounting, inventory, personnel or perhaps to control a production process? Define the nature of the software in terms of it's functions. What is it *for* exactly? Then consider the kind of user interface it employs and the type of work that users will perform with it.

B.3.7. Scope (section 7)

What is the scope of the project? Are a complete set of manuals required which detail every aspect of the software, or would it make more sense to document only those parts which people will use? Consider the benefits against the time and effort that will be needed.

The documentation needed depends on the nature of the software product, how it is applied and who will be using it. Once you've identified these, the document sets for the intended users can then be determined.

B.3.8. Topic (section 8)

Define the topic to be covered. What exactly do the users need to know? Do we want to provide background information to help them understand what's happening behind the scenes, or just tell them enough to do the job? Cost can be a major consideration here.

B.3.9. Who is the audience? (section 9)

Think about exactly who will be using the documentation. This tells you what assumptions can be safely made about how much the user already knows, and the kind of language they will understand. For example a manual which will be used by accounting staff will call for different language than a manual used by station staff.

Think about the different ways that the user will interact with the software. Will they need to interact a lot, and at what level. For example, does the software require fairly simple responses or will it need whole screens full of information to be entered? The answers to these questions determine the presentation style and the amount of detail.

In some cases, it may not be worth producing full documentation. For example a handful of highly technical users might need a relatively short manual so it would not be worthwhile producing a very detailed set of manuals in this case. On the other hand it would be worthwhile for non technical users, particularly where there are more than a handful of users.

Careful consideration of the nature of the audience allows you to choose the following:

- The best language to use.
- The right level of difficulty - not too difficult.
- How much material to include (so as not overwhelming the reader with too much or starving them with too little information).
- How to organise the concepts so that readers progress logically, beginning with what they understand and proceeding into what they don't understand.

When the manual will be widely distributed, its worthwhile keeping to the so-called "lowest common denominator" i.e. the person with the least amount of knowledge or expertise. Following this principle makes sure your manual reaches all the people it can.

B.3.10. Document set (section 10)

A set can be a single document or several documents, depending on how much detail is to be included and the needs of the audience. For example, a set of user documentation might be made up of a two volume reference manual and a training manual. The training manual is clearly separate from the reference guides which have been broken into two since to have one large volume would be unwieldy.

Where a document set needs to cater to widely differing needs, it's necessary to either include different sections for the specific audiences with that audience being clearly mentioned in the introduction, or to simply produce different documentation sets for the specific audience.

B.3.11. Presentation (section 11)

The layout and design of the document is outlined in detail in the later section "Document production". Is the document to be written according to any standard or specification?

B.3.12. Project schedule (section 12)

Develop a schedule showing the milestones, as follows:

- Blueprint.
- Research.
- First Draft.
- Language edit.
- Technical edit.
- Beta test.
- Final review date. Master copy ready.
- Manuals ready for delivery.

B.3.13. Software (section 13)

Indicate the software tools you propose to use for the project. As a minimum requirement, the various packages need to be compatible with each other. Some graphics packages, for example, won't deliver useable images to some word processors.

To avoid potential problems, stay with tried and proven software solutions or seek the opinion of knowledgeable people whose judgement you respect.

B.3.14. Editing (section 14)

Specify who will retain editorial control over the document.

Specify who will perform the technical edit.

B.3.15. Budget (section 15)

How much will the documentation project cost, or how much is being allocated for the documentation.

B.3.16. Copyright (section 16)

Who has ownership of copyright and any other proprietary rights?

In the normal course of events, the author owns the copyright. If the author has been commissioned by a second party, the formal agreement between the two parties usually prescribes that ownership remains with the commissioning party unless otherwise specified.

B.3.17. Translation (section 17)

What provision is to be made for the translation into other languages, if applicable. This includes any Localisation tools that may be used.

B.4. Estimating

Estimating the time and resources needed to successfully complete a documentation project can be a difficult procedure for the inexperienced writer. It is easy to underestimate. Estimates can be overrun if the software changes during the course of writing the manual.

The *Minutes and Hours* estimating method outlined below works on the assumption that it takes around three hours per page to write text to publication standard. The time needed to design graphics is determined by their complexity and the number of redrafts needed to ensure their technical accuracy. On average, the kind of graphic commonly found in user documentation will need three to five hours to design and amend. Graphics in this sense do not include screen images.

Knowing how long each page is likely to take does not help you to determine the likely number of pages. This needs to be done using a combination of common sense and a shrewd appraisal of the number of software functions to be documented. Your estimated page-count should be reviewed a fortnight or a month into the project.

When undertaking a large project, the deliverables should be split into manageable parts. The estimated time to complete the entire project would then be given in terms of whole months, with the first part being the only part worked out in detail.

The table below shows average times to complete the various stages of the project. It assumes the writer is entering the text directly into a PC and that desktop publishing is used.

Stage	Time
Determine deliverables	16 hrs / project
Research content	24 hrs / project
Write documentation plan	48 hrs / project
Design document structure/page layout	8 hrs / project
Write first draft	1 hr / page
Develop graphics	5 hrs / graphic
Compile text and graphics	30 min / page
Review 1st draft for technical accuracy	30 min / page
Amend draft and graphics	30 min / page
Conduct user reviews (customer)	(customer)
Incorporate user comments	30 min / page
Edit grammar	15 min / page
Prepare second draft	15 min / page
Review second draft (customer)	15 min / page

Software User Documentation

Make final corrections	10 min / page
Test documentation	15 min / page
Arrange camera-ready art	3 days
Print binders/tabs	5 days
Print and collate copies (B & W only)	10 days
Distribute	1 day

C. The writing process

This is an important section for people who are not experienced writers of non-technical prose. By applying the principles outlined here, you will extend your writing capability towards a professional level.

You may have already attained a level of competency with writing technical prose, but this is not what is required for a good user manual. User manuals describe technical matters in plain English.

C.1. Clear & effective communication

A subject as large as this could fill a library, but as big a subject as it is, there are some distilled principles which can be applied to help you write more readable user manuals.

Take the following guidelines to heart and practice them. They were largely put forward by the English writer George Orwell in his book *Politics and the English Language* (1947). Taught in university courses, practised by experienced writers everywhere, they can be considered some of the most reliable "tricks of the trade".

These principles can help you to use language as an "instrument of expressing and not for concealing or preventing thought", as Orwell said. They apply to most kinds of writing, including user manuals:

C.2. No tired figures of speech

It is a fact that when communicating, people often use expressions that have become overworked and tired (i.e. a cliché). The expression may have once had great impact, grabbing the reader's attention with the freshness of their imagery, but after 1,000 uses, they are past their "use by" date and should be retired. Take the time to think of new ways to express ideas and your writing will benefit.

As an exercise, resolve to vigilantly listen to what comes out of your mouth (and what you write) in an average day. You might be surprised at how much of it is cliché. Make a conscious effort to think of new ways of expressing yourself. Your listeners will in all likelihood appreciate the effort. You will seem more interesting to people.

C.3. Short not long words

Never use a long word when a short one will do; use "timely" not "auspicious" or "opportune. Use "set" rather than "predetermined".

It is often tempting to use a large word because we can, because we know it. We would like people to see how good our vocabulary is. Resist this temptation and use the short word. You will reach the maximum number of people with this practice.

Short words tend to be more specific or concrete, making the message more definite. Short words also usually have more impact.

Use a specific, concrete word instead of a general, abstract one. Instead of:

"We should request management to do something about their high overheads", say

"Let's ask John, Susan and Peter to suggest five ways of cutting departmental costs".

Examples of general (usually long) versus specific (mostly short) include:

- stringed instrument/guitar.
- transport vehicle/car.
- public service department/Queensland Transport.
- entertainment/movie.
- science/biology.
- sporting event/Olympic Games.

Specific words help by allowing the other person to see a clear meaning, general or abstract words tend to obscure meaning.

C.3.1. Economical & precise with words

Economical if it's possible to cut a word out without losing the meaning, always cut it out. For example to write:

"You can begin to download the data to the hard disk of the computer by loading the USB memory stick and selecting "Download" from the Utilities menu which is found in the System Administration area.."

Is not as economical as:

"To download the data, insert the USB stick and select "Download" from the Utilities Menu".

They both get the same meaning across but the first includes extra words which add nothing to the clarity of the statement, but which the reader is obliged to plough through nevertheless. In this example it isn't necessary to tell the reader where the downloaded data will go or where to insert the USB stick or even that the Utilities menu is in the System Administration area if this section is dealing with the System Administration area as a whole.

The rule of thumb is, don't make people read more than they need, the extra words get in the way, waste time and cause irritation.

Precise With around 500,000 words, English has perhaps the largest number of words of any language. With such variety, try to choose the words which best express your thought while bearing mind the advice on keeping it simple. Many words have only slight differences in meaning; i.e. *assisted,* *benefited, served, helped.* Or *meritorious, illustrious, distinguished, significant, renowned.*

The best way to achieve precision is to:

- Think carefully about what you're saying, and
- Have a broad enough vocabulary or use a Thesaurus. A good way to build your vocabulary is to make a point of looking up words you don't know and perhaps using a thesaurus when writing a document.

C.3.2. Active not passive

Always use the active voice where possible. Active voice has more impact than passive voice and is usually more concise as well. For example it's better to write:

'use the active voice'

than it is to say:

'the use of passive voice is to be discouraged'.

Notice the diluted effect that the passive voice creates. An enormous amount of what is written in organisations suffers from this problem. Why? It is partly through habit (pre-fab expressions), partly through a desire to lend an air of authority to the words and partly to hide a lack of real understanding of the subject, or worse to conceal the real meaning. Half-baked or incomplete thoughts tend to be expressed this way.

C.3.3. Everyday English not foreign, jargon or scientific

Except in situations where these are specifically called for, everyday English should be used rather than foreign, jargon or scientific words (i.e. not used for the sake of appearing knowledgeable). As a general guide, choose words that are likely to be understood by the largest number of people unless you are writing for a highly specialised readership.

It is often more difficult to use a common word when the concept is normally described in technical terms. Never assume that people know the meaning of technical words unless they have specific training (i.e. a computer science graduate can be expected to know computer jargon, but the accounts clerk who is actually using the software cannot be expected to understand computer jargon.

C.3.4. Prefabricated language

Orwell also pointed to the habit many people have of using "prefabricated" language. Rather than making the effort to think of new ways of describing things, most people lazily continue to use the same old expressions they've been using for years. For example:

'At this point, the weekly invoice run is initiated and without further ado will run until finished.'

Contains two pieces of prefabricated language; *"at this point"* and *"without further ado'*.

The result of overused expressions is that the message may not get through since the reader has tuned out after encountering too many overworked phrases. Original sounding language helps get the message across by sparking the reader's interest. In the above example, you could say:

'The weekly invoice run now commences."

Not using prefabricated language also leads to the economical expression of ideas.

C.3.5. Present tense not past/future

Unless it specifically applies, use present tense. Say *"Pressing <enter> accepts the default value"* rather than *"Pressing <enter> will accept . ."* (future tense). Another example, *"use active voice in the present tense"* rather than *"the use of passive voice in the future tense is to be discouraged'*.

Using present tense makes the message sound more immediate. The reader unconsciously thinks if it's happening now, it is worth knowing. If it's happening in the future, let's wait until it happens. If it's already happened, it's history.

C.3.6. Avoiding overstatement

This general guideline applies to all communication. While there are few opportunities for overstatement in user manuals, it's still worth mentioning. In an attempt to strengthen their message, many people resort to overstatement - words that convey an exaggerated view of a person, event or situation. If someone says "You never help me with my work" they invite a reply like "Of course I help you, what about last week?'

When a speaker exaggerates it usually makes the other person defensive - all of which gets in the way of clear communication. It's better to limit yourself to simply stating the facts; it shows that you're being fair and mindful of the other person's feelings.

C.3.7. Adapting words to the reader

To help the other person perceive what you are saying as interesting and intelligible. Certainly, using precise specific words adds interest as mentioned earlier, but you can also add interest by being concise and colourful in your phrasing.

Another way to add interest is to use colourful, non cliché expressions. For example, to describe an experience as being "electrifying" is colourful but commonplace, to say it was "like touching an electric fence" adds colour and freshness, making it both more interesting and entertaining for the listener/reader.

C.3.8. Never barbarous (advisory only)

Note: This section is for general interest only. It is included to complete Orwell's excellent list. Despite the fact that opportunities to use "barbarous" language in user manuals are very limited, it is still worth mentioning since it is perhaps the most corrupting use of language seen today.

Orwell witnessed the corrupt, dishonest use of language by World War II propagandists as a powerful but generally underestimated weapon in the arsenals of warring governments. Controlling language gives governments the means to control thought. Orwell saw not only the Nazi's doing this, but also the British, the Americans and the Soviet.

We can only think what we have the language to think with. Without language, there is little or no conceptual thought.

Orwell saw the way governments would use terms like "collateral damage" to describe the deaths of innocent people, or their own soldiers being killed by "friendly fire" (mistakenly killed by their own side), or "ethnic cleansing" for the annihilation of whole ethic groups. This language is barbarous because it disguises the reality of the situation and makes it more likely that the public will support the action.

Barbarous terms are abstract; they do not have a down-to-earth meaning. "Collateral damage" would become horrifying if the meaning was made concrete by showing the victims as real people - perhaps one's own husband, wife or children. "Ethnic cleansing" sounds almost harmless but its real meaning is barbaric when you imagine it happening in your street, to people you know.

As an exercise, listen carefully to the way governments describe their involvement in military actions. Bear in mind Rudyard Kipling's observation that *the first casualty in war is Truth.*

C.4. Non sexist language

Care should be taken to avoid non-sexist (or non-discriminatory as it is legally known) language.

As a general guide:

- **No gender assumptions** - avoid using language which assumes a person's gender. Today, there are very few jobs where a person is always male or female. Instead of saying "he/she" or "they" when mentioning a person, refer to their job title or function, i.e. "the data entry clerk" or "the user" or simply as "you".

- **Don't get carried away** with removing apparent gender bias in language. With the best of intentions it can mutilate language. For example a "manhole" cover is the generic name of the object and to call it a "personhole" cover obscures its meaning and leaves itself open to ridicule, whereas "access" cover is acceptable.

- **Further information** - if in doubt, consult the Anti-Discrimination Act and the Equal Opportunity in Public Employment Act relevant to your state.

C.5. Writer's block

Everyone has writer's block at times, even experienced writers.

Remember those times when you were trying to write an essay or assignment? You sat and stared at a blank screen or piece of paper and the words just wouldn't come. Soon you're thinking of all the things you could be doing - some of them quite important which should probably be done right away. Next thing you know, you're doing that something else and thinking "Well I'll get back to that later". This is the gentle art of procrastination whose basis lies deep in the heart of human nature.

C.5.1. Preparation

The problem is often that you're expecting to hear the finished product being dictated in your mind by that mysterious process called inspiration. But before the words will start to flow you need to know a lot about the subject. So if you are experiencing writer's block, it's generally a sign that you don't yet know enough about the subject. Spend some more time preparing and getting to know the subject well.

C.5.2. Make a start

Another tip is to lower your expectations about the quality of output at the beginning and just write what you do know even if it sounds half baked. The important thing is to start the flow of words one way or another. Concentrate on getting as much down as possible with the intention of going back and correcting it later. It doesn't matter at this stage how bad it sounds, no one else need see it. Anything you write now can be changed later in the light of a better understanding of the subject.

An analogy is that of a log jam in a river. You simply need to get the flow going again by whatever means necessary. With writing, this is by doing a "brain dump". No dynamite is required.

C.5.3. Review the reference material

If that doesn't help, go back and review the reference material you have prepared. A lack of reference material as discussed in the previous chapter is often a source of writer's block. It highlights the importance of thorough presentation to the success of the documentation.

C.6. Environment

Most people work best in a quiet, comfortable environment, as free as possible from interruptions and distractions. This is easier said than done in many workplaces, particularly when the telephone never stops ringing and co-workers frequently stop by to chat.

It is important to arrange a time and a place during the working day where you can work in a quiet, interruption free environment, since you need to be able to concentrate and follow a train of thought for an extended period. This could be early in the morning or late in the afternoon when few people are about.

C.7. Routine

Writing is a mental process that can be cultivated with regular practice. Most professional writers have a rule whereby they write a certain number of pages, or write for no less than three hours every day.

The process of writing involves using the part of your mind that performs the enormously complex task of turning ideas into language. Unless you use this acquired skill regularly, it falls into disuse and does not function well. It is similar to physical fitness. Just as regular exercise keeps a person fit, writing something every day helps keep your writing faculties in good working condition.

Schedule a period each day to work on the documentation and do everything you can to stick to the schedule. If your other commitments make it difficult to allocate time on a regular basis, discuss the matter with your manager with a view to reorganising your work schedule. It is not an unreasonable request.

C.8. Ergonomics

Since writing involves sitting in one position for long periods, certain ergonomic factors need to be considered. These include the following:

C.8.1. Chair

Provide yourself with a chair that gives good lumbar (lower back) support. Try to avoid slouching in the chair for long periods as this places strain on the lumber vertebrae, leading in time to backache.

C.8.2. Screen

The screen should be on or around eye level and not closer than around 40 centimetres. Screens emit a small amount of radiation. While no definite proof exists that this radiation is harmful to humans, many people do report degrees of discomfort and eyestrain.

The intensity of radiation coming from a screen decreases rapidly the further away the screen is. Therefore, position the screen to be as close as it needs to be to allow your eyes to comfortably read the words on the screen, and no closer.

Adjust the brightness to be just bright enough rather than brighter than necessary. If the brightness needs to be high to overcome reflected light from windows, either rearrange the screen away from the direct light, or arrange blinds. All of this helps to minimise eyestrain.

C.8.3. Regular breaks

Occupational health guidelines recommend taking a break every hour by getting up and walking around. This not only helps your circulation and eyes, it also clears the mind.

C.8.4. Keyboard

Your wrists should not need to be bent while using the keyboard. Studies show that Repetitive Strain Injury (RSI) can occur where a keyboard operator, over a long period, constantly types with bent wrists. The strain is due to the tendons which pass through the wrist from the lower arm to the hands becoming inflamed because they are being stretched and constricted as they pass through the narrow aperture in the wrist known as the Carpal Tunnel.

Avoid this possibility by making sure the keyboard is not too high. Either adjust the seat higher up, or arrange for a lower desk or a keyboard drawer which fits under the desktop, or a wrist support pad.

D.The first draft

Before beginning the first draft, make sure the preparations discussed in the first chapter have been thoroughly made.

This chapter details both the general structure of the user guide and what information needs to be included in each part. If you follow these steps, your documentation will comply with the IEEE Standard 1063 which relates to software user documentation. Include all of them unless they specifically do not apply.

Note: Try to avoid spending too much time *editing* the first draft as it is being written. It's better to write the entire first draft from start to finish, and then edit.

D.1. Backups

It's vitally important to make regular backups of your work. Without them, it's quite possible to lose days or even weeks of work in an instant and usually for no foreseeable reason. It really does happen as most readers will know! The loss can occur for a number of reasons - a software fault, a hardware fault, a power loss at the wrong time or even the theft of the PC. And it usually happens when you least expect it.

Therefore get into the habit of saving your work every few minutes. Pressing *Ctl S* does this in most WPs. Let the pressing of Ctl S become a nervous tic.

At the end of each working day, either save the document onto a USB stick, or email it to yourself as an attachment. Do both if possible.

In my own case, I *always* email the document(s) I've been working on that day to my Gmail account. When it is in the cloud it is beyond localised disasters such as a computer being stolen or a workplace burning down, destroyed by earthquake, or inundated by floodwater.

Keeping your work in the cloud also has the advantage of being able to work on it in places other than your normal place of work provided you have access to an internet connected computer.

Some employers will not be comfortable with proprietary work products being kept in the cloud. Be guided by the applicable policies in your workplace.

D.2. Deciding on the required user documents

Prepare a documentation project plan according to the details specified in Section 2 - Preparation.

Briefly this includes the following:

- Identifying the software.

- Determining the software audience.
- Determining the document set.
- Determining the document usage mode
  - Instruction mode
  - Reference mode

D.3. User document inclusion requirements

This section outlines the information required to be shown in user manuals generally. The eleven basic components of a software user document are as follows:

1. Title page
2. Restrictions
3. Warranties
4. Table of Contents
5. List of Illustrations
6. Introduction
7. Body of Document
8. Error Conditions
9. Appendixes
10. Bibliography
11. Glossary

Table 1 below shows the inclusion requirements for specific components of a document. If a component listed as mandatory contains information not applicable to a specific document, that component may be omitted (i.e. a description of conventions may not be applicable to an overview document).

Component	Single Vol. Docs		Multi Vol. Docs	
	8 Pages or less	> 8 Pages	First Volume	Other Vols)
Title page	M	M	M	M
Restrictions	M	M	M	M
Warranties	R	R	R	R
Table of contents	O	M	M	M
List of illustrations	O	O	O	O
Introduction				
Audience description	R	M	M	R
Applicability	M	M	M	M
Purpose	R	M	M	R
Document usage	R	M	M	R
Related documents	R	R	R*	R
Conventions	M	M	M	R
Problem reporting	R	M	M	R
Body				
Instruction mode	1	1	1	1
Reference mode	1	1	1	1
Error conditions	R	R	R	R
Appendixes	O	O	O	O
Bibliography	M	M	M**	M**
Glossary	M	M	M**	M**
Index	2	2	M**	M**

Table 1: User Documentation Inclusion Requirements

M = mandatory, O = optional, R = make reference, * = should address relationship in other volumes, ** = mandatory in at least one volume of the set, with references to information in other volumes. 1 = Every document has a body, 2 = index is mandatory for documents larger than 40 pages.

D.4. User document content requirements

This section describes the required content of user documents. Specific information and the level of detail for each document are determined by the audience and the usage mode of the document. The information included in this section should be included in a user document unless otherwise noted.

The titles used in this section are general and are not intended to be prescriptive; for example, "audience description" need not necessarily be labelled as such.

D.4.1. Title page

The title page of user manuals should contain the following (The cover of this document is an example of the content and format of user guide title pages):

- Document title (i.e. XYZ User Guide).
- Project name (if applicable).
- Document identification number (if applicable).
- Version number (if applicable).
- Approval information.
- Optional details for the commissioning and/or accepting of the document.
- Release date - the date the version number was last changed.

- Status - whether "draft" or "approved".
- Copy no. - controlled/uncontrolled copy.
- Company/organisation logo as specified by your organisations policy on how to display logos on printed material.

Note: Header/footer is not displayed on the title page.

D.4.2. Inside page

The page following the title page should contain:

- Document information - including the organisational unit which produced the document, the project manager's title, name, telephone and fax numbers.
- Authors name - include mention of the people involved with the preparation of the manual. Who helped? Users need to know who to consult about the manual.
- Revision history - including at least version number, date issued, author and comments regarding the purpose/updates contained in that version.
- Include mention of the software packages used to produce the manual, i.e. Word for Windows (V6.0). This may be useful information later when upgrades or additional manuals are necessary.
- Copyright notice, for example: *Copyright © Tuffley Computer Services, 2011. This publication is copyright and contains information which is the property Tuffley Computer Services. No part of this document may be copied or stored in a retrieval system without the written permission of Tuffley Computer Services.*

D.4.3. Table of contents

Note: The table of contents is the skeleton of the document. It is vitally important to properly plan the organisation of the document before creating the table of contents.

A table of contents is mandatory for every document greater than eight pages in length.

For single-volume documents this requirement should be met in one of two ways.

- Comprehensive table of contents for the whole document.

- Simplified table of contents, with a comprehensive section table of contents preceding each section.

For multi-volume documents (i.e. a single document in multiple volumes), meet this requirement by including a simple table of contents for the entire document in the first volume. In addition, each subsequent volume should contain either of the following:

- Comprehensive table of contents for the whole volume.

- Simplified table of contents for the volume, with a comprehensive section table of contents preceding each section.

D.4.3.1. Comprehensive table of contents

Construct a comprehensive table of contents for a complete document or for a section in the following way.

- Carry entries to at least the third level of the document structure hierarchy.

- Page numbers given for each entry and are joined to the entry by a leader string (row of dots) as seen in this manual's table of contents.

D.4.3.2. Simple table of contents

In a simple table of contents include at least the first level of the document hierarchy with the corresponding page number.

D.4.4. List of Illustrations

User manuals may include a list of illustrations, or separate lists for different illustration types (i.e. tables, figures etc.). If included, list(s) of illustrations should appear immediately after the contents list and should contain the following:

- Titles of all illustrations included in the document.

- The page numbers for each entry, joined to the entry by a leader string (row of dots).

Choose whether to have separate or merged lists. The type to use depends on how easy it is to understand the resultant list(s).

- Merged list - the list can be merged into one where only a few different illustration types are concerned.

- Separate list - can be used to distinguish figures, tables, screens etc. where there is a significant number of each illustration type.

D.4.5. Headers/footers

The following information should be included in the headers and footers of user manuals:

- Header contains the document ID number (refer section relating to title page), status and document version.

- Footer contains, from left to right, the project name, document title and page number.

- Both header and footer to use Arial 8pt normal.

- Header to have a narrow ruling line below and the footer a similar line above.

D.4.6. Document Introduction

The following information should be given in the introduction.

- Audience description.

- Applicability statement.

- Purpose statement.

- Document usage description.

- Related documents list or information.

- Conventions description.

- Problem reporting instructions.

The sections which follow describe these elements in detail.

D.4.6.1. Audience description

Describe the intended audience. If different sections or volumes in a set for different audiences, indicate the intended audience for each section.

In the audience description, indicate the following:

- Experience level expected of the user.

- Previous training expected of the user.

A description of the intended audience lets the read decide whether to read any further.

D.4.6.2. Applicability statement

Specify the following:

- Software version being documented.

- Hardware /operating system environment under which the software runs.

D.4.6.3. Purpose statement

Explain why the software was written and summarise the purpose of the software. Include typical intended uses of the software.

D.4.6.4. Document usage description

Describe the contents of each section, its intended use and the relationship between sections. Also provide any other directions necessary for using the document.

Example:

The manual is divided into three sections:

Section 1 Debtors Ledger used by accounts staff and management in the day to day processing of accounts receivable. Section is divided in seven sub sections which give full detail of the seven main functions.

Section 2 Creditor's Ledger used by staff in the processing of accounts payable. Section is divided in six sub sections.

Section 3 General Ledger used by staff to operate the general ledger. Section is divided in twelve sub sections.

D.4.6.5. Related documents

List related documents and give their relationship to each other. If the user documentation is made of many volumes, this "related document" information can be provided in a separate document called a "road map" or guide to document set.

Example:

Other documents relating to the XYZ Accounting Software User Guide include:

Training Guide - a set of practical hands-on exercises.

Quick Reference Guide - essential information on a laminated card.

The related documents are designed to complement not replace the user guide."

D.4.6.6. Conventions

Summarise symbols, stylistic conventions and command syntax conventions used in this manual.

- **Symbols** - describe each symbol and how they are used in the document. (i.e. "In this manual, important points are emphasised with a heavy arrow as follows: ➜")

- **Style** - explain any stylistic conventions with which the reader may be unfamiliar. These include conventions such as highlighting, boldface or italic script to indicate specific meaning. (i.e. "When a special term is first used, it is shown in boldface. Chapter names, menus and keyboard commands also appears in boldface when they are used in the context of performing an action.")

- **Command Syntax** if applicable, describe command line conventions and include examples. (i.e. "Command line entries appear in Courier boldface, i.e. `XCOPY C:\SAMPLE\*.* A: /S`")

D.4.6.7. Problem solving instructions

Give the reader instructions on the following:

- How to report software problems.

- How to contact the help desk (if applicable).

- How the reader can submit suggestions for changes in the software or the document.

List the name and contact information for the organisation responsible for responding to the problem reports or suggestions for improvement.

D.4.7. Body of document

Determine the content, organisation and presentation of the body of the document after determining whether the document is an instructional mode or reference mode document. In either mode, use a consistent organisational structure based on the expected use of the document, providing examples where necessary.

D.4.7.1. *Body of instructional documents*

Information-oriented instructional documents gives the reader background information or theory needed to understand the software. Include a scope as defined below before giving the information forming the major portion of the document. Use topics to organise an information-oriented instructional document. For example, the document could be organised as follows:

- Theory.

- Software features.

- Software architecture.

Task oriented instructional documents gives the reader the necessary information to do a specific task or attain a specific goal. Provide all of the information listed below relating to scope, materials, preparations, cautions/warnings, method and related information. Use task relations to organise a task-oriented document or section (i.e. Organise by task groups or task sequence).

Scope. Begin this section by indicating to the user the scope of the material to be discussed.

Materials. Describe any materials the user will need to complete the task (i.e. input manuals, passwords, computers, peripherals, cabling, software drivers, interfaces and protocols). Alternatively, describe separately the materials common to all or many functions and refer to that description.

Preparations. Describe any action, technical or administrative, that should be done before starting the task (i.e. obtain system passwords, access authorisation, disk space). Alternatively, describe in a separate section the preparations common to all or many functions and refer to that section.

Cautions and Warnings. Describe general cautions and warnings that apply to the task. Place specific cautions and warnings on the same page and immediately before the action that requires the caution or warning.

Method. Describe each task, including the following:

- What the user should do

- What function, if any, is invoked, (including how to invoke the function and how to recognise normal termination).

- Possible errors, how to avoid them and how to solve them.

- What results to expect.

Related Information Provide other useful information about the task, including the following:

- Tasks frequently done together and their relationship.

- Other tasks customarily performed by the users of this document that could be supported by the methods of this section. Describe this support.

- Notes, limitations or constraints (notes may also be placed in the specific area to which they apply).

D.4.7.2. *Body of reference documents*

Organise a reference mode document the way a user accesses a software function. Methods include the following:

- By command

- By menu selection

- By system calls

Within this organisation, arrange the functions for easy access and random user access (i.e. alphabetical order or a menu-tree hierarchy). For each function, include all of the information listed below relating to purpose, materials, preparations, inputs, cautions/warnings, invocation, suspension of operations, termination of operations, output(s) error conditions and related information.

Purpose. Describe the purpose of the function.

Materials. Describe any materials the user will need to complete the task (i.e. input manuals, passwords, computers, peripherals, cabling, software drivers, interfaces and protocols). Alternatively, describe separately the materials common to all or many functions and refer to that section.

Preparations. Describe any action, technical or administrative, that should be done before starting the task (i.e. Obtain system passwords, access authorisation, disk space). Alternatively, describe in a separate section the preparations common to all or many functions and refer to that section.

Input(s). Identify and describe all data required for the correct processing of each function. Use one of the following methods.

- Describe inputs used only by a single function in the section devoted to that function.

- Describe in a single section or in an appendix inputs used by multiple functions. Refer to that section or appendix when describing these functions.

Cautions and Warnings. Describe general cautions and warnings that apply to the task. Place specific cautions or warnings on the same page and immediately before the action that requires the caution or warning.

Invocation. Provide all information needed to use and control the function. Describe all parameters. Include the following:

- Required parameters

- Optional parameters

- Default options

- Order and syntax

Suspend Operations. Describe how to interrupt the function during execution and how to restart it.

Termination of Operation. Describe how to recognise function termination, including abnormal terminations.

Output(s). Describe the results of executing the function, for example.

- Screen display.

- Effect on data or files.

- Completion status values or output parameters

- Outputs that trigger other actions (i.e. mechanical actions in process control application.)

Provide a complete results description for each function. If several results are possible, explain the situations that produce each.

Error Conditions. Describe common error conditions that could occur as a result of executing the function, and describe how to detect that the error has occurred. For example, list any error messages displayed by the system. (Error recovery need not be included here if they are covered in the later section "Error messages, known problems etc'; - see below)

Related Information. Provide other useful information about the task, including the following:

- Tasks frequently performed together and their relationship.

- Other tasks customarily performed by users of this document that could be supported by the methods of this section. Describe this support.

- Notes, limitations or constraints (notes may also be placed in the specific area to which they apply).

D.4.8. Error messages, known problems & error recovery

Provide a list of error messages in a convenient location, i.e. in a separate section, chapter or appendix.

A complete error message index includes the following:

- A list of each error message and/or code with an explanation for each.

- The error that caused it.

- How to recover from it.

- The action(s) needed to clear it.

- Describe known software problems here or in a separate document and provide alternative methods or recovery procedures.

D.4.9. Appendixes

Include in appendixes any material which supports or augments the information given in the main part of the manual. The list that follows is exhaustive and some points may not apply except in highly complex and/or technical situations.

- Input and output data or formats which are used in common by a group of functions.

- Input and output codes (i.e. inventory codes).

- Interactions between tasks or functions - i.e. a description, illustrated with diagrams of the relationships and interaction between groups of tasks or functions.

- Global processing limitations if applicable.

- Description of data formats and file structures.

- Sample files, reports or programs.

D.4.9.1. Bibliography

If applicable, provide a list of any publications that may have been quoted or referenced in the manual. Where applicable, other sources containing related information can also be included to point a curious reader in the direction of further information.

The bibliographic entry includes:

- Author's name and initials.

- Year of publication

- Title in italics.

- Publisher's name.

- The second and subsequent lines are indented.

Sample format (APA):

Herzog, F. 1989. *Systems Analysis & Design*. Brisbane: University of Queensland Press

D.4.10. Glossary

List alphabetically in the glossary definitions of the following:

- All terms, acronyms and abbreviations used in the manual with which the read may be unfamiliar.

- All terms, acronyms and abbreviations with which the reader may be unfamiliar (i.e. the way in which they are used or their context).

D.4.11. Index

Development an index, based on key words and concepts, for any user document greater than eight pages in length.

Construct the index as follows:

- Indicate the importance of information, place minor keywords under major ones. (i.e. "Printer" could be further divided into cabling, connection to, error messages and fonts).

- Give the page location to the right of each entry.

- List location references in one of the following ways:

  o By page number

  o By section or paragraph number

  o By illustration number

    o By other index entry

- Use only one level of index entry. Where an entry points to a second index entry, the second entry should give the location and not point to a third entry.

D.4.12. User documentation presentation requirements

The document presentation requirements are outlined in detail in section 5 of this guide.

The following are general guidelines:

D.4.12.1. Consistency

Use terminology and typographic and stylistic conventions consistently throughout the document or set of documents. Identify any deviations the first time they appear.

D.4.12.2. Terminology

Define all terms requiring a glossary entry, acronyms and abbreviations when they appear for the first time. This means the glossary appears twice; once in a section of it's own and again distributed throughout the text as the term is used for the first time.

D.4.12.3. Referencing related material

If related material is placed in separate parts of the document, or in separate documents of a set, repetition of the information can be avoided by providing specific references to the related information.

D.4.13. Change control

D.4.13.1. This-version function changes

Software changes made while documentation is being written which need to be reflected in the documentation. You need a process by which changes in the functionality of the software being documented are made known to you so that they may be reflected in the documentation.

This procedure often involves the documenter receiving a copy of the software change control form which details the nature of any changes made.

D.4.13.2. Next-version function changes

Software functionality changes made while the documentation is being written but which are not to be reflected in the documentation on publication. The changes will be reflected in subsequent publications.

The distinction between "this-version" and "next-version" is usually made on the basis of a cut-off date - i.e. any changes after an agreed date will go into the next version.

E. Document production

This chapter outlines the recommended document production standards for user manuals. It is suggested that you use the document template supplied with this manual (*userman_template.docx*), downloadable from:

http://www.tuffley.com/template/userman_template.docx

It is recommended that you save and rename this file as your working draft manual.

This template provides the basis for your documentation project. It conforms to the details specified in the sections below.

E.1. Document orientation

All material in the user guide with the exception of appendixes should be presented in "portrait" format. Mixed orientation in the body of the document simply confuses the user, and requires constant turning of the document to read the contents.

Copies of reports output by a system should be included in appendixes and so may be produced in "landscape" format due to their width.

E.2. Document organisation

The user manual should be organised according to the order outlined in the earlier section called "User document inclusion requirements" (Section 4).

E.3. Document format

This section outlines a uniform set of formatting standards which enable a reader to find the information they need with a minimum of time and effort.

E.3.1. Double sided

User guides should be produced as double sided documents from a single sided master. This reduces the bulk of the user reference guide, and makes the user reference guide easier to read as this adopts the normal "open book" format familiar to most readers.

When this format is used, it makes it a simple process to display input screens on the left-hand pages, with the relevant instructions for the input required in each field on the right-hand pages. This greatly assists the user when they are referring to the user reference guide, or using it as a training aid for new staff.

E.3.2. Table of contents

The document contents list is located after the preface page and should comply with the following:

- Be titled "Contents" not "Table of contents".

- Contain the headings from the document broken down and arranged into a logical groupings.

- Contain all headings up to but not exceeding level four headings. The document itself should have no more than four levels of headings.

- Illustrations lists should be shown.

- Appendix(es) should be shown.

- Page numbers are given for each entry and are joined to the entry by a leader string (row of dots).

- TOC text should use the following font attributes:

Level	Font	Indent	Size	Space above/below
toc 1	Arial bold	1.0 cm.	12 pt.	12/6 pts.
toc 2	Arial	1.5 cm.	11 pt.	6/0 pts.
toc 3	Arial	2.0 cm.	11 pt.	3/0 pts.
toc 3	Arial	2.5 cm.	11 pt.	0/0 pts.

- Level one TOC entry should have a narrow ruling line beneath.

- Header/footer.

If the document being produced does not have a table of contents, it is good practice to design the document before beginning work by making a list of the headings which will form the contents list.

As a general guide, the organisation of a document should reflect a logical sequence, whether a chronological sequence, a progression from the general to the specific, or other meaningful sequence.

E.3.3. Page size and margins

Page set-up should have the following attributes:

- Page size - A4.

- Margins (left & right) - 2.54 cm.

- Margins (top & bottom) - 3.17 cm.

- Header & footer - 1.25 cm in from edge.

- Gutter - zero cm.

E.3.4. Headings

Document headings should have the following attributes:

- No more than four levels of heading should be used.

- Headings should be indented according to their level of significance in accordance with the following table.

Lev	Style	Font & Size	Space above/below	Ruling line	Indent
1	Heading 1	Arial 18 pt bold	18/18 pts.	3 pt.	0 cm.
2	Heading 2	Arial 16 pt bold	10/12 pts.	1.5 pt.	1.25 cm.
3	Heading 3	Arial 14 pt bold	12/6 pts.	-	2.5 cm.
4	Heading 4	Arial 12 pt bold	12/3 pts	-	3.75 cm.

Level one and two headings should have a ruling line beneath, as shown in the above table. The ruling lines should be 4 pts. beneath the text and extend from the beginning of the heading across to the right margin.

All headings should have the first word capitalised and subsequent words in lower case. Proper nouns (i.e. Queensland Rail) have the first letter capitalised wherever they appear in a heading.

Note: This guide species the use of Arial and Times Roman fonts since these are excellent general purpose fonts available on most computers. As a general rule, stick to these two fonts where possible. If unavailable use another similar font. Avoid the use of unusual or fancy fonts unless you have a specific need.

E.3.5. Body text

The body text should have the following attributes:

- Be left justified (i.e. aligned to the left, leaving a ragged right edge).

- Times Roman (TT) 12 points.

- Indented 3.75 cm from the left margin.

- Space - zero points above and six points below.

E.3.6. Bullets

Bullets are used for lists where the order is not important. Bullet points should have the following attributes:

- Times Roman (TT) 11 points.

- Indented 3.75 cm from the left margin so that bullet character is in line with body text.

- Bullet character a 10 pt. filled dot.

- Hanging indent of 0.5 cm from bullet character to start of text.

- Space - zero points above and three points below.

- Bullet points should only be used for lists of two or more points (i.e. no single bullet point).

E.3.7. Representing screens

E.3.7.1. Overview

Generally, putting screens into your manual is to be encouraged, but the decision to include them will be influenced by several factors:

- **Expertise with computers** - the reader's expertise with computers and software. Screen images are good where the reader is not comfortable or familiar with computers. Screens are helpful in showing the reader what they should be seeing. Where the reader is highly experienced, screens are less important unless;

- **Specific Information** - there is specific information on the screen about which the reader should be aware. In this case the same information needs to be highlighted in the main text as well as the screen.

- **Maintainability** - manuals with screens usually need more maintenance than those without. That's because those without are less likely to be affected when changes are made to the software. Think about how much time and effort is likely to be available for maintenance. Since inaccuracies like incorrect screens are a quick way for people to lose confidence in a manual, it's better to leave screens out if there's not enough time to maintain them properly.

E.3.7.2. Screen capture techniques

If you do decide to include screens, you need either:

- **Screen capture** - a screen capture program which takes a snapshot of whatever is on the screen at a given time and saves it as a file which can then be imported into your document. There are several different products on the market which do this, ranging from inexpensive shareware up to sophisticated products that allow the file to be saved in a wide variety of formats.

- **Windows** - if you're familiar with Windows and the software to be documented is capable of running in a window, a utility is available on the menu that appears when the button at the extreme top left corner of the window in which the application is running is pressed. Select the option to "Edit" from this menu, then "Mark Area" to mark out with the mouse cursor the area to copy. The marked area is now in the clipboard and can be pasted into Notepad where it can be saved as an ASCII file.

Screens that are saved as images are usually much larger than those saved in ASCII format. The document quickly becomes very large (i.e.. over 100Kb) when more than a few images are included in it. So saving the file in ASCII format is recommended since it takes up less space.

E.3.7.3. Screen requirements

Screens include data input screens, enquiry screens, and job submission screens.

The information displayed in screens should be meaningful but not refer to actual people or situations. Likewise, no sensitive information should be displayed.

Screens should have the following features:

- Be indented so that the left edge lines up with the body text, and it should have ruling line around.

- No shading - it is not necessary to shade the screen to give it more of a "screen-like" appearance.

- A non-proportional font like courier should be used to represent screen text where possible.

- The screen should be labelled at the bottom left corner of the box. This label should be highlighted in bold, and show the business function name followed by the screen name. For example: **Transfer Menu: Set up a Transfer - Transfer Details**

- Screens should not carry over onto the next page.

E.3.8. Keyboard function keys

Keyboard function keys (including program function keys and keyboard control keys) should be represented in upper case bold enclosed within angle brackets. For example:

- <ENTER>

- <CTRL>

- <PF1>

- <END>

- <HOME>

E.3.9. Numbered lists

Numbered lists are used where the order is significant, such as in procedural steps.

The attributes for a numbered list are the same as for bullet lists, with the substitution of Arabic numbers (i.e. 1, 2, 3) for bullet dots and a hanging indent of 0.75 cm after the number to accommodate double digits.

E.3.10. Figures & numerals

Decimal numbers less than 10 shown as numbers (i.e. 6.5). Whole numbers less than 10 are spelt in full. Spaces are used for numbers with four or more figures (i.e. 10 000).

E.3.11. Justification

Justification styles should be as follows:

- All heading, body text and bulleted/numbered lists and captions left justified.

- Table headings and table contents may be left, right or centred.

E.3.12. Italics

Italics should be used for the following:

- To add *emphasis* to words, phrases or sentences.

- Foreign words (i.e. *ad hoc*).

- Titles of publications (i.e. *Concise Oxford Dictionary*)

E.3.13. Emphasis

Emphasis can be achieved in the following ways (in ascending order of emphasis):

- Italics.

- Bold Italics.

- Callout character - as follows:

Note: Add emphasis to a paragraph with a callout character. Words within such a paragraph may be further emphasised with *italics* or ***bold italics***.

E.3.14. Hyphenation

Hyphenation is not obligatory. Many people prefer not to have body text hyphenated since it slightly increases the difficulty with which a reader scans a line, although the difficulty in many cases is negligible.

Where done, hyphenation should be made as follows:

- The break is made after a vowel with the second part beginning with a consonant (i.e. peo-ple).

- For multiple part words the practice of hyphenating words is decreasing, so as a general rule don't hyphenate if in doubt. For example, cooperate multi tasking.

- Fractions are hyphenated (i.e. two-thirds, five-eighths). Compound words like day-to-day task-sharing are hyphenated.

Enable automatic hyphenation where this is available in your word processor.

E.3.15. Acronyms & jargon

While the use of acronyms and jargon can be justified in some cases, they should be avoided where possible. Where acronyms are used they should be defined in the definitions and acronyms section and also defined the first time they are used in each chapter.

Only acronyms used in the current document are to appear in the definitions & acronyms section.

E.3.16. Widow/orphan protection

A "widow" is the last line of a paragraph that appears alone at the top of a page. An "orphan" is the first line of a paragraph that appears alone at the bottom of a page.

Use the widow/orphan protection option in your word processor to prevent this occurring in the document.

E.3.17. Paragraph numbering

While headings can be numbered, avoid paragraph numbering the body text unless there is a specific need to use it.

E.3.18. Other considerations

Abbreviations - use fullstops with abbreviations and contractions. For example Fri. for Friday.

Punctuation Spacing - one space only after any punctuation mark.

F.   Editing & proof-reading

This chapter outlines what to do after you've written the first draft based on the guidelines in the previous chapter.

F.1. Editing

The task of editing begins after the first draft is complete.

When formatting the first draft, use the principles outlined in the previous chapter as a guide.

F.1.1. Print first draft

The first draft is generally entered into a word processor and the result formatted and printed using either:

- The word processor's own desktop publishing abilities.

- A desktop publishing product to accept the text produced by a word processor.

Note: For some time now, there has been convergence between WP and DTP whereby all but the most demanding top-end publishing jobs can be accomplished with a word processor. For high-end jobs, Framemaker is a solid choice, though there are others.

The choice is influenced by what is the standard in your organisation. You are normally bound to follow the standard unless a compelling reason not to is given.

If you intend using a DTP product, do not format the text as it is typed into the word processor. Any formatting that is included will need to be removed or altered when it's imported into the DTP product.

F.1.2. Binding

Once the first draft is formatted and printed, put it in a binder. This is useful because:

- Organising the material makes editing easier.

- It gives you a preview of the finished product.

F.1.3. Edit first draft

Editing marks a turning point in your attitude towards the document. It calls for a shift from a subjective to an objective point of view. You need to shift your perspective from being the one who wrote it to the one who is now going to criticise it.

While nobody likes having their work criticised, being your own harshest critic raises the standard of your work above average and so helps to create a professional result.

Imagine you are the reader, then look for:

- **Clarity.** Whether the manual can be understood. This includes whether the language is appropriate for the reader, the information is organised logically and clearly and the layout helps not hinders the reader to find information.

- **Brevity.** Ways to be more concise. The first draft can always be made more concise, since the object has been to get the information down first, then worry about polishing it later. The reader will appreciate your efforts here. Think of how frustrating it is when you've been made to read an entire paragraph or page of convoluted double-talk when the real message could have been said in a sentence. It's a waste of everyone's time.

- **Active voice, present tense.** As mentioned earlier, active voice present tense language has more impact and immediacy, and that's good because it helps to get the message across.

- **Sexist language.** Check for unwarranted gender assumptions and language which may cause offence due to gender bias.

- **Illustrations & Examples**. Are there enough of both to support and clarify the information being presented?

Editing is time consuming when done properly.

F.1.4. Edit second draft

After making the changes arising out of the first edit, print
and bind it in the same way as above. While the first edit
looks for improvements to the language, format and content,
the second edit looks for technical accuracy and consistency
of presentation.

Ask yourself "Have I delivered what was promised?".

F.1.5. Spell/grammar check, TOC, index

After implementing the changes from the second edit, do a
thorough spellcheck, and if you have it available, also a
grammar check. Grammar checking programs allow for
different kinds of writing (i.e. general, legal, business,
technical etc.) Make sure you select the right category.

Also generate the table of contents and the index.

F.1.6. First proof

After doing the checks and generations mentioned above,
print and bind the result. It is now ready to be proof-read.

F.1.7. Checklist

- **Acronyms**. Check that uncommon acronyms are written out in full the first time they appear in each chapter.

- **Bulleted Lists.** Check for correct indentation and consistent bullet size both within and between lists.

- **File Names, Path Names, Operator Entries.** Check for consistent use of typeface.

- **Headers and Footers.** Make sure section and chapter titles in headers and footers are correct

- **Headings.** Check for proper and consistent capitalisation. Also check for consistent phrasing of headings.

- **Index.** Check accuracy of all references by looking them up in the manual.

- **Key Names**. Check for consistent use of typeface and abbreviations (i.e., Ctrl for Control).

- **Measurements**. Make sure units of measurement are used consistently (i.e., all SI units followed by imperial units in parentheses).

- **Numbered Lists.** Check for correct numbering sequence; look for repeated, missing, or out of order numbers.

- **Page Numbers.** Make sure pages are in the correct order.

- **References**. Check for consistency of typeface and phrasing when referring to other publications and other parts of the manual. Check that references to page numbers, figures, and tables are correct.

- **Spelling**. Run your electronic spell checker AND check manually (neither method is accurate enough by itself).

- **Symbols**. Make sure symbols are used consistently (i.e., if you use a symbol such as a square to mark the end of a numbered list, make sure it appears at the end of every list).

- **Table of Contents**. Make sure page number listings match page numbers in the manual.

- **Word Choice**. Make sure wording of instructions is consistent (i.e., do not use "choose" and "select" to mean the same thing; use one or the other).

F.2. Proof-reading

Proof-reading begins after the editing is complete. From a practical point of view, the manual is as finished as you can make it without being checked by a third party.

F.2.1. Who proof-reads?

Finding the right people to proof-read your almost complete manual is the final challenge to overcome. Print out four copies of the manual, place them in binders and write "Draft: dd/mm/yy. Copy No. n of 4" on the cover to identify it as a draft produced on a certain date and the copy number out of the total of four.

The process happens on these four levels:

- **Accuracy** - the information given in the manual is checked for technical accuracy. Approach the technical people who developed the software with your most polite request for help. Most of the time they aren't obliged to help and many technical people aren't thrilled by the prospect of proof-reading a manual. It's often useful to talk to the technical manager first and ask who might be able to help. Then see the nominated person and begin by saying "I was talking to (your boss) and he suggested I see you about having a look at this manual . . .". Explain that it is the technical detail you'd like them to check. Spelling and punctuation will be checked by someone else.

- **Spelling & Grammar -** while the word processor's spell/grammar checker is a useful first pass, it's still advisable to have a literate person check the manual for spelling and grammar. In addition to general checking, ask them to look for unnecessary use of passive voice and past/future tense.

- **Usability** - the manual is checked for usability by the same people who will use it. As with the technical proof, begin by talking to a manager in the user area and explain that for their own benefit you'd really appreciate it if someone could check the manual from the user's point of view. This can also be done as part of the "Acceptance Testing" stage of the software development.

- **Says too much** - the manual is checked for content which might inadvertently give offence or disclose information of a sensitive nature.

F.2.2. Deadlines

Make it clear to the proof-readers that their help is appreciated. At the same time ask whether a month (or whatever, depending on the size of the manual and your own timeframe) will be long enough. If their reply is along the lines of :

- **"it should be"** say you'll get back to them in three weeks or so. Always give the proof-reader a specific deadline and then diplomatically check with them a week or so before the deadline.

- **"I don't think so"**, ask how long they think it will take. If that's too long (remembering it almost always takes people longer than they say) then consider asking someone else.

F.2.2.1. Make the changes

While most suggested changes should be implemented, there will be some that don't, in your opinion, improve the manual. If this happens try to resolve it with the person concerned, otherwise ask a third party who has the authority to settle the matter. Remember, whatever the input from other people, it's you that takes the ultimate responsibility for the manual.

G. Reviewing & field testing

This chapter deals with the important subject of getting useful feedback about the usefulness of the document from people who are qualified to comment.

Why do we need to do these reviews and tests? The main reason is that no-one, not even the most selfless among us, are completely ego-less when evaluating their own work. Since we don't want to find fault with our work, we tend to overlook faults when we see them. In other words, documenters who see their manuals as extensions of their egos are not going to try and find all the errors in their manuals.

G.1. Reviewing the manual

As ego less documenters we need to include a review stage so that the end result will better suit the user.

Guidelines for conducting reviews:

- Select reviewers and the time when the review will take place.

- Show the reviewers ho to conduct the review.

- Give feedback to the reviewers.

G.1.1. Selecting reviewers and when to review

More often than not, it is necessary to find several different people who can bring a different perspective to a document. That is unless you are lucky to have a person who can adopt multiple perspectives when reviewing a document.

The following perspectives need to be brought to bear:

- **Technical perspective** - checking for technical accuracy and completeness.

- **Management perspective** - checking for how well the document achieves it's overall objectives, how well it projects a positive image of the organisation and how well it protects proprietary information.

- **Editorial perspective** - checking for conformity with accepted writing conventions.

- **User perspective** - checking the document for it's "usability".

The matter of whether to circulate a separate copy to each reviewer or give each reviewer the same copy, each taking their turn, can be tricky. While it saves time to circulate multiple copies, you run the risk of being caught in the middle between two reviewers who disagree on the same point. Giving each reviewer the same copy, one at a time, might take longer but it reduces the risk of conflict.

Where a conflict exists, arrange for a general review meeting and ask them to resolve the issue.

G.1.2. Show reviewers how to review

The reviewers need to be given specific guidelines on how to conduct an effective review. Unless people have had specific training in this field, they are unlikely to know how to do it properly, since it does not feature in school or university curriculums.

As a general rule, the quality and quantity of the reviews you receive are directly related to the amount of time and effort you spend spelling out what is needed from the reviewers.

The first step is to convene and initial meeting, or circulate a initiation memorandum. At the meeting or in the memo, you should outline the general picture: what your purpose is, who the audience is, what special techniques your manual uses, etc.

One major problem which often occurs is that the reviewers confuse which perspective they should be using. The technical reviewer might slip into editorial perspective, while those looking from a management perspective can sometimes do the work of the technical reviewer.

A good way to instruct the reviewer is to place the instructions in the form of a checklist on the coversheet. The checklist could be along the lines of the following:

- Is the purpose of the material clear and accurate?

- Is the definition of the audience's needs and experience complete and accurate?

- Do the graphics suit the audience?

- Has all required information been provided?

For each review type (technical, management etc.) a separate checklist in addition to the one above should be included. For example, with the management review:

- Does the information show accurately the benefits of the system?

- Can all promises made in the document be kept?

- Does the material protect the confidentiality of our organisation proprietary information?

As a final courtesy, when the document has been finalised, send a copy to each of the reviewers with your compliments and thanks.

G.1.3. Give feedback to reviewers

Regardless of how much time and effort a reviewer has spent, they should all be given feedback. Let each know the changes that will or won't be made based on their review comments. Feedback will increase the chances that the person will agree to review other documents in the future.

G.2. Field testing the manual

In field testing, users of the documentation are given the chance to use it and test it's ability to stand alone. The aim is to let real world experiences improve the quality of the finished product.

Guidelines for conducting a field test are as follows:

- Select testers and the time when the test will take place.

- Show the testers how to conduct the review.

- Give feedback to the testers.

G.2.1. Selecting testers and when to test

When the manual is close to looking like the finished product. Don't wait until the document is finished, it needs to be field tested before final release.

Field testing takes a minimum of three weeks and may take six to eight weeks for complex systems

G.2.2. Show testers how to test

Since you won't be there to personally instruct the testers, you need to develop an explicit set of instruction for them to use.

The testers need to be prepared psychologically for their task by means of a letter that contains the kind of material discussed in the previous section on reviewing. For example you need to take the time to explain to the testers the general picture, what the purpose is, etc.

in addition, a coversheet should accompany the documentation. The coversheet should contain the following:

- Identify the document as a field test document that is to be used for evaluation purposes only and not for final release.

- Explain the purpose of the test.

- Explain how long the test will take, and indicate the document must be returned to you at the conclusion of the test in order to avoid the document from "living on" past their useful lives.

- Explain how changes and suggestions should be made - on the draft copy, in a meeting, over the phone or whatever.

G.2.3. Give feedback to testers

Just as it is important to make the reviewers feel appreciated, it is likewise so for the field testers.

Let each know the changes that will or won't be made based on their test results. Feedback will increase the chances that the person will agree to test other documents in the future.

H.Production

This section explains the "production" or printing stage. It assumes your manual is black and white since colour is both relatively rare and expensive for most user manuals.

H.1. Paper size & weight

Size User manuals should be printed on white A4 paper, preferably on both sides. This is regarded as the most practical size for presentation, ease of reading, and binding. Since it is a standard size, the masters do not need to be photo reduced before production, thus improving the manual's "readability". It can also be drilled and bound in standard A4 binders, and is a size familiar to the users.

Weight Pages should be copied or printed on 110 gsm paper. 110 gsm is relatively heavy duty and is particularly good when the ring-binding method is used, since pages can quite easily be torn out of a ring-binder by a heavy-handed or perhaps frustrated user. In general though, this is the most practical and durable weight for any printed matter that is handled frequently.

A blank buffer sheet of 150 gsm paper can be used as the first and last page of the user reference guide. This prevents the photocopied pages adhering to the plastic covers of the binders.

H.2. Print quality

The master copy of the manual should be printed on at least a 300 dpi, preferably 600 dpi laser printer. The master should be checked for degree of blackness, smudging, and paper quality. It is imperative the master is of sufficiently high quality to copy successfully.

H.3. Binding

User reference documentation should be bound in loose leaf four ring binders. These binders should be white plastic with clear plastic insert pockets on the front and spine. This means the document can be easily updated, and additional pages or chapters slotted in later at minimal cost. It allows binders to be reused as required.

(Four ring binders have been selected as the standard, as the contents are less easily torn out or accidentally damaged, than in a two or three ring binder.)

H.4. Drilling

All material included in any user reference guide, or subsequent update, should be drilled prior to distribution. All pages are to be drilled on the left-hand side in the standard position suitable for four ring binders.

These holes will be drilled no less than 1.5 centimetres from the left-hand edge of the paper.

H.5. Tab dividers

Each chapter of the user reference guide should be separated by a tab divider that clearly indicates the end of one chapter, and start of the next.

Ideally these dividers should be clearly labelled with the name of the chapter that immediately follows it, as it is listed in the master table of contents.

These tab dividers should be made of a durable material such as plastic. This ensures they do not accidentally tear out, and can be gripped to open the user reference guide at the section required.

H.6. Registration

A Project Document Register should be established as soon as the project is commenced. All project documents (including external project documents, correspondence, memos, etc, but excluding Email) should be recorded in the project document register.

Project documents from external sources should be given an appropriate document identification number to allow them to be placed in the project filing system.

Project documents should be listed in ascending order of document number, within document subject, within document type.

The Project Document Register should contain the following information for all project documents.

- Document identification number.

- Document title.

- Document location - the full directory path and filename of the working copy of documents which exist as soft copies.
  The words "Paper only" (or similar) for documents which exist as a hard copy only.

- Document version - the version number of the latest edition of the document.

The Project Document Register should enable staff to perform the following:

- Allocate the next available document number within a given subject to a new document and record the allocation of the number to ensure each document number is unique.

- Locate any project-related document.

The Project Document Register should be implemented by one of the following methods:

- A hard copy document.

- A software application such as a document table, spreadsheet, or database program.

Note: Where there is the likelihood that unauthorised (therefore unregistered) copies of a manual will be made, one safeguard is to inscribe the serial number using a red marker pen. Since red does not reproduce well in photocopiers, it will at least be apparent by looking at the cover of a manual whether it is an unauthorised copy.

H.7. Identification of changes

As changes are made to a user guide and a new version of that document is produced, the changes should be indicated in the revision history record. The information should be sufficiently detailed to identify each area of change.

Optionally, the changed paragraphs may be marked with revision marks. New text will be underlined and text to be deleted will be marked with strike-through characters.

H.8. Document release

Approved or unapproved documents are released to users, project team members and others on a need-to-have basis. In other words you issue as many as necessary, but no more. The more documents issued, the harder it is to control them with regard to making sure everyone gets updates.

The project manager authorises the release of project documents.

H.9. Version control & re-issue of documents

Documents which are subject to version control, and which are issued as controlled hard copy, should be issued in one of two ways.

- **The total document is re-issued** each time a new version is produced. This is the most straight forward approach, and only requires the recipient to dispose of the earlier version and commence use of the new version. The revision history record need only deal with the version number of the total document and the changes in each version.

- **Only updated pages are re-issued**. This requires the maintenance of version control at the page level, and the recipient should remove and replace the revised pages. It is more difficult for the user to identify if they are reading the most recent version of a page. The revision history should deal with individual pages. Where done, the additional page version control information is inserted before the table of contents.

The choice of method should be made depending on the type and size of the document, and the frequency of issue of new versions. If at all possible, the re-issue of the total document should be considered.

A means of avoiding these problems is to distribute the document as controlled soft copy only, and if the document is in hard copy form it is uncontrolled.

For controlled distribution documents, refer to the section of this Standard relating to controlled distribution.

H.10.Controlled distribution documents

In rare cases, a user guide will contain secret and/or highly sensitive information which require them to be strictly controlled. The following conditions are mandatory for controlled documents.

- Where the document has "controlled" hard copies, those people/positions who are in receipt of a "controlled" copy should be known by means of a register.

- Where a new version of a document is distributed, it should be accompanied by a notice to the relevant staff which identifies the reason for the change and the procedure for them to update their existing copy.

- The new version and the notice need go only to the holders of "controlled" copies.

I. On-line documentation

On-line documentation is becoming more common due to the availability of easy to use on-line document generators and the increasing trend towards networking computers together.

Until recently, on-line documents - technically known as "hypertext" documents - could only be produced with difficulty, by a process not unlike computer programming.

Various options for on-line delivery, whether via the web or intranet means that it is now viable to convert paper documents into hypertext documents, place them on the network and make them available to all. This approach has some obvious advantages. Hypertext documents don't need to be printed or physically transported to the user, both of which saves time and money. And due to the ease with which an on-line document can be copied into a LAN directory, it's now possible to ensure that the latest version of a document is the one being used.

Another reason on-line documentation is becoming more common is that several of the biggest problems with early on-line documents have been solved. Early on-line documents used to be screen after screen of characters, often at poor resolution which made them difficult to read, and with no effective way of knowing where you were within the document as a whole. These days, the resolution of computer screens is much improved, which solves the first problem.

I.1. Conversion to on-line

The master copy of a manual should still be paper-based. From this paper-document however, it's then possible with some but not all hypertext generators to create the on-line document.

It's preferable to use the one document as both the master and the basis upon which the on-line version is made.

The guiding principle with on-line documentation is that it will not altogether replace paper-based documentation, but it does have an important part to play in the delivery of up-to-date user reference information.

I.2. Create the paper manual first

Whether you intend using on-line documentation or not, it is still necessary to first create the paper version. Having done this, you can then convert it to an on-line version using a suitable product. The on-line version should change little, if at all from the original.

The point is, the on-line document always happens second. The majority of the work you do will be in the creation of the paper-based master copy of the manual.

J. Maintenance

If you've followed the steps outlined in this book, your manual should be an effective piece of user documentation, and as such is worth maintaining so that users will continue to have confidence in it.

J.1. When to maintain

When to maintain a manual:

- **Add and/or amend details** - when information has been added or changed. For example, in a ledger system, the account codes might change and new ones be added. When a significant number of such changes have accumulated, the updated pages are sent out together rather than singly.

- **Clarify information** - when the information needs clarification in order for the reader to understand. Likewise accumulate these changes as above.

J.2. Procedure

How to maintain a manual:

- **Edit manual progressively** - edit the manual as the changes happen. If you leave to accumulate for several months, it's often difficult to find enough time to do them.

- **Update Log** - maintain a record of all changes made to a manual, including the date, details of the change and the reason for the change.

- **Print updated pages** and place them in a folder or envelope.

- **Distribute updated pages** - prepare a covering letter explaining the nature and scope of the enclosed updates and forward a copy of everything to each person on the distribution list.

J.3. Change control

There need to be a process by which changes in the functionality of the software after the documentation has been written are made known to you so that they may be incorporated in future editions of the documentation.

Two kinds of change control are recognised.

J.3.1. Post-publication software changes

Software changes made after the documentation has been published

J.3.2. Post-publication document changes

Changes to the published documentation which derive from either software changes or the rectification of software "bugs".

In both cases, the required procedure often involves the documenter receiving a copy of the software change control form which details the nature of any changes made.

K. Glossary of computer terms

---

access time

The time needed to find information stored in a computer.

acoustic coupler

An electronic device which enables information from a computer to be transmitted across a telephone line by converting the computer's digital output to an acoustic signal.

alphanumeric

Letters from the alphabet and/or the numerals 0 to 9.

APL

A programming language generally used for advanced mathematical work. [from A Programming Language]

applications package

A purpose written set of programs - for example payroll, ledgers. The package is made up of several programs - see next entry.

applications program

One of several programs making up an applications package. An applications program performs a particular job, like calculating long service leave entitlements in a payroll package. There would be many applications programs needed to make up a complete payroll package.

applications study

Determining the computer software and hardware needed for a particular purpose.

architecture

The overall design of a computer system. Similar in nature to the architecture of a building in that it shows the relationship between the elements that make up the whole.

archive

When used as a noun, the store of data transferred by the process of archiving. As a verb, refers to the act of transferring data to a suitable storage medium (usually magnetic tape or diskette) once it is no longer needed for day to day purposes.

artificial intelligence

Programs which allows computers to make decisions in apparently the same way humans do. This includes learning by trial and error, understanding natural language and interpreting visual and auditory signals. AI makes a machine act like a human. AI systems have been developed to give medical and legal advice.

ASCII

American Standard Code for Information Interchange. The commonly agreed standard for encoding instructions into seven bit binary code, allowing different computers to share information.

assembly language

A low level language which uses a form of notation that a computer interprets as "machine code" but which the programmer can readily understand.

assembly program

A program to converts assembly language into machine language.

attach

Connect a device to a system or network so that it's resources are available to other users or devices on that system/network.

audit trail

A record of every action performed by a computer. Generally used to detect unauthorised access to information and/or trace faults.

backup

When used as a noun, the store of data transferred by the process of backing up. As a verb, refers to the act of duplicating data to a suitable storage medium (usually magnetic tape or diskette) so that it can be restored if the original is lost through accident or misfortune.

band printer

An impact line printer which has the font etched on a steel band. Generally used where high speed printing is necessary.

bar code reader

An input device for the optical scanning of bar codes. The information is converted into the kind of digital signals which are understood by computers. This is a quick and error free way of entering information into a computer.

BASIC

Beginner's All purpose Symbolic Instruction Code - a simple and popular programming language for personal computers

batch

To collect a group of tasks into a batch for easy handling by a computer.

batch processing

Processing a quantity of similar information by grouping it together as a unit before inputting it into the computer. Compare REAL TIME.

baud

The unit for measuring the rate at which electronic data is transferred between computers. One baud equals one bit (of information) per second. Many modems transfer at 9600 baud, which means 9600 bits per second are transferred along a telephone line. Named after French engineer J.M.E. Baudet.

benchmark

A standard by which the performance of a computer system is measured. For example, a benchmark is often a program which is run on a selection of machines and the time taken to complete as well as the quality of the results is used to compare the various machines with each other.

binary digit

see BIT.

bit

The most basic unit of information in a computer, represented as 0 or 1. [from Binary digit]

bit mapping

A method of displaying graphics where e each pixel corresponds to one or more bits in the processor's memory.

BNF

Backus Normal Form (or Backup Naur form) - a the formal notation describing the syntax of a programming language.

Boolean Algebra

A kind of mathematical logic used in computer programming. Named after mathematician, George Boole).

boot

To start a computer by loading it's operating system then any applications into memory.

bootstrap

The program which loads the operating system automatically upon switching on (or booting the system). Derives from the expression "to haul oneself up by the bootstraps" where a person hanging upside down climbs up his own bootlaces to get free.

bps

Bits per second - a unit for measuring the rate at which electronic data is transferred between computers. See also BAUD.

browse

Searching through computer files for information not specifically known to exist.

bug

An often undetected error in a computer program.

bus

Parallel channels which carry information from one part of a computer system to another. For example a 32 bit bus carries 32 bits of information in parallel.

byte

An arrangement of bits (normally eight) which represent a single character and are treated as a whole by the computer.

C

A high level computer language which allows the implementation of the UNIX operating system and which also allows the programmer to avoid using assembly language.

cache memory

Memory used to temporarily store data from a the hard disk. Because cache memory can be accessed more quickly, this allows for improved performance.

CAD

Computer aided design - a software program which produces plans and drawings similar to those produced manually by a draftsman or engineer.

central processing unit

CPU - the main operating part of a computer. Includes the arithmetic, logic and control units.

chad

The confetti like material produced when printer paper is drawn across the tractor sprockets of a printer's paper feed mechanism.

character

An element of a given character set or a subdivision of a word.

chip

A small piece of semiconductor material on which an integrated circuit is printed. The entire chip forms a functional circuit.

clock

An electronic pulse generator, the frequency of which controls the speed of a computer system.

COBOL

COmmon Business Oriented Language - a programming language developed early in the history of computing for commercial applications.

code

A piece of programming language.

cold boot

To restart a computer by manually switching the power off and then back on. Contrast WARM BOOT.

concurrency

Executing two or more processes at the same time on the same machine.

COM

Computer Output on Microfilm - making microfilm directly from magnetic tape or signals. Also COMPUTER ORIGINATED MICROFILM; MICROMATION.

compiler

A computer program which translates high level programming languages into a low level one.

computer originated microfilm

See COM.

computer output on microfilm

See COM

control character

A character which when typed at the keyboard initiates a sequence of events (or control functions). Control characters are not treated as normal data.

corruption

Loss of data contained in a storage medium such as a diskette, hard disk or magnetic tape

CPU

See Central Processing Unit

crash

A system failure which requiring human intervention to remedy.

cursor

The reference point visible on a screen to indicate the current position.

daisy wheel printer

A kind of impact printer in which the font is formed on the end of spring fingers that extend radially from a central hub. The daisy wheel can be easily exchanged to enable alternate character sets to be used. The name derives from it's resemblance to a daisy flower.

data

The general name given to information as entered, stored and/or processed by a computer.

databank

see DATABASE

database

a large volume of information stored in a computer and organised in categories to facilitate retrieval.

any large collection of information or reference material. Also
DATABANK.

database management system

see DBMS

data capture

Collecting data to be used by a computer. Usually refers to
automatic data capture, for example recording sales data by a
cash register linked with a central computer at another
location.

data dictionary

The structured description of the entities contained in a
specific database, as distinct from raw data.

data entry

The process of entering information into a computer system.

data processing

Using a computer to do routine operations involving the
storage and processing of data; for example the posting of
transactions in a ledger. DBMS

An integrated DataBase Management System with the ability
to define the logical and physical structure of the data
contained in a database and for accessing, entering and
deleting data.

debug

To find and correct BUGS in a computer program. The term
originated in the United States in the early days of computing

when a system failure at a large installation was caused by a moth becoming stuck in a relay switch and the action taken to rectify the problem was recorded in the log as "debugging".

Desktop Publishing (DTP)

A kind of software package which allows the sophisticated formatting and production of documents on relatively inexpensive personal computers. Until it's widespread introduction in the 1980's, the same result was available only through a Printer or Typesetter at a much higher cost.

DTP has revolutionised the publishing industry, reducing production times from days to hours and cutting production costs proportionately.

diagnostic routine

A program within a program which tries to determine the cause of a software or hardware error.

digital computer

A computer which processes digital data.

directory

A list of the files stored on a floppy disk or other storage device. 2. One of the storage areas into which a hard disk may be divided. The root directory gives the user access to the other directories (first level directories), which in turn may have subdirectories of their own.

disk

A data storage device usually a thin, magnetically coated disc. A disk is either removable (floppy disk or diskette) or non removable (HARD DISK).

disk drive

The transport mechanism which rotates a disk (see previous entry) in order for read/write heads to retrieve or store information.

diskette

see DISK.

disk operating system

see DOS.

DOS

Disk Operating System - an operating system developed by the Microsoft Corporation for IBM. It became the standard operating system on the first personal computers in the 1970s. In the 1990's DOS had become the pre eminent operating system for PC's. It's future past the mid 1990's is uncertain in view of Microsoft's moves to develop more efficient operating systems for the highly competitive PC market.

dot matrix printer

Type of printer in which individual characters are made with ink dots from a rectangular matrix of printing positions.

download

The transfer of programs and/or data from a mainframe "down" to other smaller computers. Reverse process is upload.

down time

The time in which a computer system is unavailable for use.

dumb terminal

A relatively simple terminal with no capacity for storing or handling data. It is generally used to input information into and display information from a mainframe computer.

dump

As a noun, a magnetic tape record which duplicates the contents of a disk or a printout of the contents of a computer's memory at the moment a problem has occurred. As a verb, to make a magnetic or printed copy.

EFTPOS

Electronic Funds Transfer at Point Of Sale. See ELECTRONIC FUNDS TRANSFER.

electronic funds transfer

The transfer of funds electronically from a debtor's (customer) bank account to a creditor's (shop) account via a computer terminal installed at the point of sale. Abbrev.: EFT.

electronic mail

Messages sent between users connected to a computer system.

expansion slot

A vacant position inside a computer into which an additional circuit board (also known as a "card') may be placed.

father file

see FILE.

field

A subdivision of a record, comprising a number of characters which are treated as a whole.

fifth generation

The next generation of computers which will be designed to have human like intelligence.

file

An organised collection of related records. Grandfather, father and son files are the three most recent versions of a periodically updated file.

flag

A data storage character which indicates whether a certain condition exists or not after a logical operation has been performed.

floppy disk

see DISK.

flow chart

A diagram indicating the structure of a program or algorithm and the sequence in which operations are performed.

font

A complete set of printing characters of one style and size, for example Courier or Times Roman.

foreground/background processing

Sharing processor time such that specific tasks are given priority and are processed by the computer on demand, while other less important tasks are processed in the remaining time.

FORTRAN

FORmula TRANslation - a computer language for scientific calculations in which instructions are expressed in an algebraic notation.

friction feed

Mechanism for advancing paper in printers by gripping it between rollers.

GIGO

Garbage In Garbage Out - a truism which maintains that a program working with incorrect data produces incorrect results.

global

A variable which is accessible from all parts of a program.

golf ball printer

Type of impact printer with the type mounted on a spherical, golf ball shaped print head. Derives from the original IBM golf ball typewriter.

hacker

An unauthorised person who seeks to gain entry to secure areas of a computer system.

handshake

The control device which allows data to be transferred across an interface.

hard copy

A printed copy of data as opposed to an image on a screen.

hard disk

see DISK

hardware

the physical components of a computer system, such as the CPU, disk drives, keyboard, screen, printer, etc.

high level language

A type of programming language similar to human language which is then translated into machine code which the computer can understand.

hit

The object of a search. When a file is searched for a particular item, the number of occurrences found is the number of hits.

impact printer

A type of printer which uses the same printing action as a conventional typewriter; that is, characters being pressed through an inked ribbon onto paper. Used in daisy wheel, golf ball and dot matrix printers.

ink jet printer

A type of printer which produces characters by spraying a fine jet of quick drying ink in the shape of the text characters.

integrated circuit (IC)

An electronic circuit whose highly miniaturised components are contained on a single chip of semi-conductor, allowing a very complex circuit to be produced in a relatively small package.

intelligent terminal

A type of terminal with software that allows a certain amount of computing to be done without contact with a central computer.

interface

The point of contact between two systems.

interrupt

a command to switch from one program, usually one which is running in the background, to another, and then back again. 2. such a suspension of program. 3. of or relating to a system of interrupt.

joystick

An input device which allows the cursor position to be controlled. Mainly used with recreational software. Derives from the name given to the control device of an aeroplane.

K

approximately a thousand; 2^{10} words, bytes or bits. 2. a kilobyte. 3. a thousand of some unit, as in a salary package of 60K.

KB

A kilobyte; 1024 bytes. Also K.

key

The record identifier used in information retrieval.

keyboard

The familiar data entry device consisting of a board with rows of keys. A QWERTY keyboard is laid out in the standard typewriter pattern. A separate additional numerical keypad is often provided on computers to speed numerical data entry.

key in

To enter data into a computer via a keyboard

key to disk/tape encoder

an input device for accepting data from a keyboard and writing it directly to magnetic disk or tape.

LAN

Local area network of computer systems, eg linking computers on a university campus via cables.

laser printer

Type of printer which uses a laser beam to form the characters on paper.

LCD

Liquid crystal display.

LED

Light emitting diode.

light pen

A light sensitive device in the shape of a pen used to interact with a computer by shining a point of light on the screen.

line printer

Type of printer which prints a complete line at a time, often at rates which may be thousands of lines a minute (contrast with machines which print a character at a time).

low level language

Machine oriented language in which each program instruction corresponds to a single machine code instruction. Assembly language and machine language constitute two distinct categories of low level language.

low resolution graphics

Graphical display units which can construct pictures by plotting relatively large blocks of colour or by using special graphics characters. Contrast HIGH RESOLUTION GRAPHICS.

machine code

MACHINE LANGUAGE

machine language

Low level language composed of complex binary codes which is a precise set of operating instructions for a computer, as opposed to a more symbolic, generalised code or a high level language. Also MACHINE CODE.

machine readable

Of or relating to data that can be input to a computer without further preparation, such as magnetic ink characters.

magnetic disk

see DISK.

magnetic ink character recognition

see MICR

magnetic tape

A plastic tape coated with a ferromagnetic powder, esp. iron oxide, used to record digital information in computing, and machine instructions in industrial and other control systems (as well as to record sound in tape recorders and video signals in video recorders).

magnetic tape encoder

An input device that accepts data from a keyboard and writes it directly to magnetic tape.

media

Material (collectively) for storing data, incl. disk, tape, paper, cards, etc.

memory

The part of a digital computer in which data and instructions are held until they are required.

menu

a range of options presented to an operator by a computer. A particular choice may itself lead to a further menu.

the screen displaying the menu, or the file where it is stored.

pull down menu, one available to the operator at any time, when requested, obscuring part of the screen. After a choice has been made, the pull down menu disappears and the screen returns to its previous state.

MICR

Magnetic ink character recognition; a machine reading system by which ferrous impregnated ink characters encoded on documents such as cheques are read by a magnetically sensitive device.

modem

A device enabling connection of peripheral devices to a computer, esp. one converting bits into analogue electrical impulses for transmission over the telephone lines. [from MOdulation/DEModulator]

monochrome

(of a VDU) featuring different shades of a single colour.

mother board

A printed circuit board that holds the principal components in a microcomputer system.

mouse

Manually activated pointing device for positioning the cursor on a VDU.

multiprocessor system

A computer with two or more processors; thus being faster.

network

An arrangement of connected computer systems and peripheral devices. Ring network and star network are two types of network.

OCR

optical character recognition; a system of machine reading by a light sensitive electrical cell of (a) standard character sets encoded on documents such as cheques or (b) one of a number of standard typefaces. The number of typefaces able to be read by OCR are increasing.

optical character reader; a machine (scanner) able to recognise printed characters.

OEM

Original equipment manufacturer; an organisation which produces computer hardware for other manufacturers to build into their products.

optical character reader

See OCR.

optical character recognition

See OCR.

overwrite

To write data to a disk location which is already occupied by existing data. This existing data is lost in the process.

package

SOFTWARE PACKAGE.

parallel interface

An interface that communicates eight bits at a time. Compare SERIAL INTERFACE.

parallel printer

Type of printer with a parallel interface (as opposed to a serial interface).

parallel processing

Computer processing of two or more tasks simultaneously. The computer may require more than one processor to do this (i.e. a MULTIPROCESSOR SYSTEM).

PASCAL

A programming language developed to encourage structured programming. Named after Blaise Pascal.

peripheral

A device connected to a computer which either transfers data to or from the central processing unit of the computer. Peripherals include printers, modems, etc.).

personal computer

A microcomputer used by an individual, usually for business purposes rather than as a home or hobby computer. Abbrev.: PC.

pixel

The smallest element of a graphics display - a single dot.

PL/1

A programming language used in both scientific and business computing. [from Programming Language 1]

point of sale terminal

See ELECTRONIC FUNDS TRANSFER.

port

A point in a circuit at which an external connection is made. The point at which a peripheral is connected.

portability

In relation to software, the ability to work effectively on a variety of computers, with a variety of compilers, or with a variety of operating systems.

printed circuit board

The insulated board forming the base onto which computer components are soldered and connected as a circuit; commonly abbreviated to BOARD.

printer

Device for producing a printed copy of data.

processor

General word for a data processor or CPU.

prompt

Character or symbol displayed on a VDU indicating that the user is expected to input data or carry out some operation.

pull down menu

See MENU.

QWERTY keyboard

See KEYBOARD.

RAM

Random access memory. Temporary memory structured so that each item can be accessed quickly though information is stored only while the computer is turned on.

random access memory

See RAM.

read only memory

See ROM.

real time

Information processing dealing with data which has been just entered (as in airline ticket reservations) as opposed to batch processing which accumulates data and processes it as a group at a later time.

record

A self contained group of data, treated as a unit for processing. For example a complete creditor record, including creditor number, address, credit status, etc. In a fixed length record the number of characters is predetermined; in a variable length record it is not.

report generator

A program which generates reports from one or more files, usually by prompting the user to specify what files and data to use, what operations to carry out on the data, the desired report format, etc.

reset

to re boot (a computer).

to restore (an indicator) to an initial state, eg a flag to show whether the most recent version is backed up.

ring network

See NETWORK.

ROM

Read only memory; permanent computer memory that can be read but not altered or erased by program instructions.

root directory

See DIRECTORY.

RS 232C

A standard interface for serial data communication.

second generation

see COMPUTER GENERATIONS

sector

the smallest addressable part of a track on a magnetic disk or tape, commonly 128 or 256 bytes.

the block of data stored on a sector.

serial interface

An interface that communicates information one bit at a time. Compare PARALLEL INTERFACE.

serial printer

A printer with a serial interface.

software

A computer program or the collection of programs which cause a computer to perform a desired operation or series of operations (as opposed to HARDWARE). "If it can be physically touched it's hardware; if not, it's software.'

software house

An organisation that manufactures and markets software as programs and packages.

software package

A fully developed program(s) to perform a specified set of tasks, such as ledgers, sales order processing.

son file

See FILE.

spooling

The temporary storage of data on disk or tape to free a computer for further processing, since a normal printer prints data at a rate much slower than the computer can send it.

spreadsheet

A software product designed for financial and/or statistical processing. Well known spreadsheets include Lotus 123 and Microsoft Excel.

subdirectory

See DIRECTORY.

subprogram

A part of a program which performs a specific task. Also SUBROUTINE.

subroutine

SUBPROGRAM.

system board

The board containing the central processing unit.

terminal

An input/output device connected to a computer but at a distance from it. It is generally composed of a keyboard and a VDU.

thermal printer

A printer which uses heat sensitive paper (thermal paper), to produce characters by the action of heated wires.

third generation

See COMPUTER GENERATIONS.

time sharing

The provision of access to a computer for two or more users at the same time (from their individual terminals). Each user is allowed alternating time slices of the system's resources, esp. the CPU, thus appearing to have continuous use of the system.

track

The band or path on a tape or disk along which data is stored. On a tape the tracks are parallel; on a disk they are like the tracks on a vinyl record.

tractor feed

The mechanism for advancing continuous stationery on a printer which uses toothed wheels (sprockets) to grip the perforations along the edges of the paper and feed it forwards.

upload

See DOWNLOAD.

utility program

A program designed to carry out fundamental functions, such as listing and copying of files, sorting data, etc. Also UTILITY SOFTWARE; UTILITY.

variable length record

See RECORD

VDT

Visual display terminal. See VDU.

VDU

A visual display unit; a terminal device with a cathode ray tube on which text and images are displayed. It is usually used in conjunction with a keyboard. Also VDT.

videodisk

A storage medium primarily used for videofilm. The very high storage capacity of videodisks makes them attractive as a future storage medium for computer data.

visual display unit

See VDU.

warm boot

The restarting of a program without first switching off the system (as opposed to a COLD BOOT). Also WARM START.

window

a part of the VDU screen, such as a text window in a graphics screen, or the area taken up by a pull down menu.

that portion of a large document currently visible on the VDU screen.

word processor

A software program allowing text entry, formatting, manipulation and printing.

write enable notch

The slot on a diskette allowing the disk to be used.

write protection

Protecting a storage medium, usually a diskette of tape, from accidental erasure. Involves setting a write enable tab.

L. Bibliography

Bowen, J. 1983. *Pocket Guide to the Law in the State of Queensland.* Sydney: Law Foundation of NSW.

Brockmann, R. J. 1990. *Writing Better User Documentation.* New York. Wiley & Sons.

Crystal, David. 1987. *The Cambridge Encyclopaedia of Language.* Cambridge: Press Syndicate of the University of Cambridge.

Dear, I.C.B. 1987. *Oxford English: A Guide to the Language.* Oxford: Oxford University Press

Holmes, Lubelski, Powell and Ranger. 1989. *Desktop Publishing Design Basics.* Blueprint Publishing, London.

International Organisation for Standardisation. 1985. ISO6592 *Information processing - Guidelines for the documentation of computer-based application systems.*

Institute of Electrical and Electronics Engineers. 1993. *IEEE Standards Collection, Software Engineering,* 1993 Ed. New York: IEEE.

Lane, Leroy. 1991. *By All Means Communicate.* New York: Prentice Hall.

Orwell, George. 1947. *Politics and the English Language.* London.

Rudwick, Jennifer. 1989. *The Macquarie Office Manual 2nd Ed.* Sydney: The Macquarie Library.

Standards Australia. 1994. *AS/NZS 4258:1994 - Software User Documentation Process.* Sydney.

M.   Appendix A - Forms

Appendix A provides a series of forms to assist in the advancement of a documentation project.

They include the following:

- Project overview form.
- Project personnel form.
- Project schedule.
- Documentation set description.
- Documentation description.
- Audience profile.
- Provisional table of contents.

Make photocopies of the forms and maintain these as your master copy.

Project Overview

Project name _____

Brief description of project_____

_____

_____

Why was this project initiated? _____

_____

_____

_____

_____

_____

Brief description of user interface design _____

_____

_____

_____

_____

_____

_____

Other comments about project _____

_____

_____

_____

Documentation Project Personnel

Document Name _____

Manager(s) _____ Phone _____

_____ Phone _____

Programmer(s) _____ Phone _____

_____ Phone _____

Writer(s) _____ Phone _____

_____ Phone _____

Reviewers _____ Phone _____

_____ Phone _____

_____ Phone _____

_____ Phone _____

_____ Phone _____

Editor _____ Phone _____

Proof-reader _____ Phone _____

Software User Documentation

Word processor(s) _____ Phone _____

_____ Phone _____

Artist _____ Phone _____

Production supervisor _____ Phone _____

Scheduler _____ Phone _____

Other(s) _____ Phone _____

_____ Phone _____

_____ Phone _____

_____ Phone _____

Documentation Schedule

Document name _____

Blueprint _____/_____/_____ Note _____

Research _____/_____/_____ Note _____

First Draft _____/_____/_____ Note _____

Language Edit_____/_____/_____ Note _____

Technical Edit _____/_____/_____ Note _____

Beta Test _____/_____/_____ Note _____

Final review date _____/_____/_____

Master-copy ready _____

Books complete/ready for delivery _____/_____/_____

Notes _____

_____

_____

Documentation Set Description

Project name _____

Printed Documents

_____ No. of pages _____

_____ No. of pages _____

_____ No. of pages _____

_____ No. of pages _____

_____ No. of pages _____

_____ No. of pages _____

_____ No. of pages _____

On-line materials

_____ No. of screens _____

_____ No. of screens _____

_____ No. of screens _____

_____ No. of screens _____

Packaging notes

_____

_____

_____

_____

Documentation Description

Complete a form for each printed piece of documentation.

Title of document _____

Estimated page length

Production Information:

How will pages of document be reproduced? _____
(Photocopy, offset printing, other)

How will text be created? _____
(Word processing program, test editor, other)

How will typeset pages be produced? _____
_____
(Laser printer, commercial typesetter, imageset)

How will pages be made up? _____
(Desktop publishing, traditional paste-up, other)

What types of illustrations will be used? _____
_____
(Line art, halftone art, photographs, screen shots, tables)

How will document be bound?

Ring binder _____ Size _____

No. Of rings _____

Saddle stitching _____

Wiro binding _____ Size _____

Perfect binding _____ Soft or hard cover __

Other _____

Page layout:

Attach document specifications and sample pages, if available.

Page size:_____

Number of heading levels _____ Sideheads _____

Columns:

Multiple-column tables _____

Two-column procedures _____

Two-column lists _____

Software User Documentation

Special type styles (bold italics, full caps, small caps, contrasting type style):

Element   Format

_____   _____

_____   _____

_____   _____

_____   _____

_____   _____

_____   _____

_____   _____

_____   _____

What types of illustrations can the writer use? _____
_____

(Diagrams, concept drawings, photographs, screen shots, tables, charts, graphs)

Maximum size illustrations _____

Special graphic treatments (boxes, screen tints, second-colour ink, lines and rules, other):

Element   Treatment

Audience Profile

Job title(s) _____

_____

_____

Knowledge of computer hardware and software that these users probably already have:

_____

_____

_____

Reasons these people might buy this product _____

_____

_____

_____

Features that would especially interest these people _____

_____

_____

_____

Secondary Group:

Job title(s) _____

_____

_____

Knowledge of computer hardware and software that these users probably already have:

_____

_____

_____

Reasons these people might buy this product _____

_____

_____

_____

Features that would especially interest these people _____

_____

Provisional Table Of Contents

Introduction

Chapter 1   _____

_____

_____

_____

_____

Chapter 2   _____

_____

_____

_____

Chapter 3   _____

_____

_____

_____

_____

Appendix   _____

Index   _____